N I G E L C A M E R O N

THE ROBOTS ARE COMING: US, THEM & GOD

NIGEL M. DE S. CAMERON

Nigel Cameron writes about the ethics and policy impacts of new technologies. A Scot who has lived many years in the United States, he has taught in universities, founded a think tank in Washington, and written widely for both religious and secular markets. Together with Joni Eareckson Tada, he wrote *How to be a Christian in a Brave New World* (Zondervan, 2006). His most recent book is *Will Robots Take Your Job? A Plea for Consensus* (Polity, 2017).

Published by CARE

ISBN 978 0 905 195 23 0

British Library Cataloguing-in Publication Data
A catalogue record for this book is available from the British Library

CARE 53 Romney Street London SW1P 3RF
020 7233 0455
mail@care.org.uk
care.org.uk
Charity No: 1066963
Scottish Charity No: SC038911

Design: Dave Potter | david-potter.co.uk

Generic images: shutterstock.com, istockphoto.com & fotolia.com

THIS ONE FOR MY DAUGHTER ALICE – WHO LIKE ME IS A NON-TECHNOLOGIST WORKING WITH TECHNOLOGY EVERY DAY.

O Lord, our Lord, how majestic is your name in all the earth!
You have set your glory above the heavens.
Out of the mouth of babies and infants, you have established strength because of your foes, to still the enemy and the avenger.
When I look at your heavens, the work of your fingers, the moon and the stars, which you have set in place,
what is man that you are mindful of him, and the son of man that you care for him?
Yet you have made him a little lower than the heavenly beings
and crowned him with glory and honour.
You have given him dominion over the works of your hands;
you have put all things under his feet, all sheep and oxen,
and also the beasts of the field,
the birds of the heavens, and the fish of the sea,
whatever passes along the paths of the seas.
O Lord, our Lord,
how majestic is your name in all the earth!

Psalm 8 1-8

In an age of ever increasing technological change Nigel Cameron unpacks in an accessible way what these developments could mean for us as individual Christians, and for the Church as a whole. In the quest 'to make robots human', what is beneficial to us and what is dangerous? What does it mean to be human in the twenty-first century? These questions are handled thoughtfully by Nigel, giving us even more food for thought.

Nola Leach
Chief Executive, CARE

CONTENTS

PREFACE

When Nigel first suggested that CARE should consider the future of robotics and the ethical and practical consequences they might have on what it means to be human, I confess I was not enthusiastic. But as he unpacked the subject I began to realise that the astonishing advances being made in artificial intelligence and computer technology will transform society. Already machines are at work in factories where people used to be employed. Friendly humanoids are caring for the elderly in Japan. Apple's driverless trucks are revving up to navigate US highways. Realistic interactive sex dolls are big business.

What does it all mean for us? Should we welcome this latest industrial revolution or fear it? The future of work and leisure is at stake. We will also face questions about the status of humanoid machines – do they have rights? Should they be treated as if they were one of us? Because we care about the dignity of human beings created in God's image, these questions are important for Christians to consider.

This fascinating subject is new to many of us and in this stimulating book Nigel introduces us very clearly to the issues involved, from a biblical perspective. He takes us on a journey which starts with the earliest 'robots' and, in an Epilogue, into the future into an imagined world of 2040.

As Christians we need to get our heads around the ethical and practical possibilities of the coming of robots and this book, and the questions and issues is raises, is an excellent place to start.

Rev Lyndon Bowring
Chairman, CARE
August 2017

INTRODUCTION

ARE THE ROBOTS REALLY COMING?

The world will soon be teeming with new creatures. It will be the most dramatic change in the history of the human race. It promises to be wonderful, and to be terrible, but above all to be confusing. Because these life-forms will be made by us. They won't be people. They won't be animals. But also they won't be 'things' in the sense in which we have understood things in the past. We don't yet know much about them and how they will develop. But we can be sure of some facts:

- they are developing very fast
- they are smart – and will keep on getting smarter
- they will take on more and more of the tasks that used to be our responsibility
- they will work for us, and alongside us
- and they will become more like us all the time.

Christians believe that God made the universe, and God made us. We believe He gave us dominion over the rest of His creation. He put us in charge – to use our minds and our hands for Him. He gave us responsibility for the birds and the beasts and the fish – but also for everything else on the earth. Our God-given intelligence has used the raw materials of which this earth is full to invent everything from the wheel and the plough to cars and aeroplanes.

Since the middle of the twentieth century we've witnessed something new, as the speed of invention has moved at a faster and faster pace. Someone has calculated that our modern computers are **one trillion** times as powerful as the most powerful computers in the world around the year 1960. And those miniature super-computers we call mobiles or smartphones are themselves **one million** times more powerful than the massive old-time mainframe machines available to NASA when they put the first man on the Moon in 1969. One million times more powerful. It's worth repeating, as it's hard to grasp what it means. And the reason we need to try is that the process that has led to these enormous gains in our ability to make powerful computers hasn't stopped.

Every single day that passes, our ability to make these super-intelligent machines even smarter increases. And they get smaller. And cheaper. Every day.

As they get smarter, they become just a little more like us. They have a long way to go before they can do everything we humans can do, but they can already do some bits better than we can. For example, using a satnav to navigate is already a lot simpler and easier than having someone give you directions. Even if that someone is sitting next to you in the car reading a map!

What are we to make of all this? It isn't that on a special day – say, 12 August 2018 – someone is going to invent a Robot, and we shall have to decide whether we approve or not. Robots of many kinds have been taking over aspects of our lives for years. Their impact is getting bigger, they are becoming more intrusive, and they are also becoming more helpful. Day by day. It's been a gradual process and although it keeps speeding up it is still gradual, step-by-step, one thing at a time.

But there are obvious challenges and problems and future impacts that are easy to see – and that we need to start talking about. What about using Robots to take care of our older people? And to act as nannies to our children? What about Robots making decisions that have huge results – like driving cars and lorries that could be involved in accidents? What about Robots designed for sex? Or, more subtly, Robots designed to be friends and companions and take the place of humans?

And what about jobs? Already Robots are driving trains that people used to be paid to drive, and dispensing cash from machines when we used to wait for the bank to open. Are they going to take so many jobs that people find it harder to get work?

These are mostly rather complicated questions. Try talking about some of them with your friends and you will realise that we disagree on our answers, whether we are Christian believers, Jews or Muslims or have no faith at all. Is there a Christian view of Robots?

Are there Christian perspectives that we need to feed into public discussion, as the Robot revolution arrives? Are there ways in which Christian believers should choose differently from other people – whether the issue is Robot toys and 'friends' or Robot sex or Robots as nannies or carers and companions for our old people?

And what about government? Governments find technology issues difficult to handle, partly as technology moves fast and governments almost always move very slowly. Partly because governments tend to act when they sense that people feel strongly about something – like cloning, for example, or genetically-modified foods – but on an issue like Robots people don't quite know what to think. So governments are likely to sit back.

Of course, governments love technology 'innovation' as it spells good for the economy. But Robot innovations aren't quite like others.

One reason why innovation has been seen as such a good thing is that it brings jobs. But today's high-tech companies are different. They aren't hiring tens of thousands of people to make cars and televisions and aeroplanes. They actually use a lot of Robots, even if Robots aren't what they are making and selling; and Robot technologies are designed *precisely not to create jobs*. And the more high-tech these companies are, the less interested they seem to be in hiring people to work for them.

It's also true that politicians are very reluctant to raise problems about new technology. Who wants to be labelled a Luddite? That's also true of business leaders and even of trade unionists. They all want to be seen as on the side of progress and the future and welcoming tomorrow. Luddite really is a bad word.

These are all reasons why you are unlikely to hear many speeches by politicians on Robots and why they are wonderful, or terrible, or whatever blend of the two they might decide.

And when did you last hear a sermon on Robots? You are pretty unusual if you have ever heard even one. Perhaps your church leader makes the occasional disapproving comment on how we are addicted to our mobiles – especially if one goes off in church! But beyond that?

In fact, church leaders of all kinds tend to be wary of discussing science and technology. It's partly because they don't feel like 'experts', and they suspect someone out there in the congregation knows a lot more than they do. Especially if they are a little older and they know that younger people – teenagers and student-age especially – know far more about the latest technology. Understandably, they don't want to say things that will make them look stupid.

There are various reasons why I have written this little book. One of them is to help raise the level of understanding of what's going on in the world of technology – which is right at the top of current developments. If you are a 'techie' you track all these things, and plenty who would not see themselves as techies make the effort to read magazines like *Wired*, online or in print, and keep up. But most people – including most church leaders – don't, and there's not much use just telling them that they should.

I'm hopeful that Christian churches and other congregations of people of faith will be able to take a lead in helping the rest of us think through these remarkable developments – as they keep ramping up into a future that will be dominated by Robots. Because we need to keep our eyes fixed on the importance of humans taking care of humans, whatever wonderful machines we use in the process. I want a human future, with people at the centre. I hope you do too.

Nigel M. de S. Cameron

CHAPTER ONE

SPOOKY STORIES

'I do think, in time, people will have, sort of, relationships with certain kinds of robots – not every robot, but certain kinds of robots – where they might feel that it is a sort of friendship, but it's going to be of a robot-human kind.' **Cynthia Breazeal, Robot engineer**

We're getting used to being entertained by spooky stories about Robots behaving like people, and real people treating them as if they were people.

In the 2002 film *S1M0NE*, Al Pacino plays a director who is in trouble as his star has pulled out. So he creates one of his own using his computer. He calls her S1M0NE, which stands for 'Simulation One'. The film is such a success – S1M0NE (actually played by Canadian actress Rachel Roberts) becomes an instant star – that Pacino's character Viktor decides to launch a whole career for 'her'.

On a similar theme, the 2013 film *HER* tells how the lead character, played by Joaquin Phoenix, in the midst of a divorce, 'falls in love' with the computer-generated voice of actress Scarlett Johansson. Once you've seen the film, it's hard to get it out of your head when you next hear Siri or Cortana or your satnav 'voice' giving you driving directions.

BICENTENNIAL MAN

Perhaps the most thoughtful Robot film is *Bicentennial Man*. Made back in 1999, it stars Robin Williams as a household Robot named Andrew, and is based on a short story by famed sci-fi writer Isaac Asimov.

At first Andrew is thoroughly clunky. But as time passes he keeps being upgraded, until he looks and sounds very like a human being. In the process – surprise! – one of the humans in the household falls in love with 'him', and 'he' falls in love back.

That's a very quick summary! There is a lot going on in this film, including a world court that is asked to decide whether Andrew can be regarded as a human being. The first decision is no, as Andrew does not age and humans all age. So he works on turning himself into a being who ages like humans do – and therefore who will ultimately die.

The court then changes its mind, recognises him as a human and therefore someone who is able to get married. And then, now a very old Robot who is visibly ageing, Andrew does finally die.

AI: ARTIFICIAL INTELLIGENCE

One of the best films to explore the relations of Robots and people was Stephen Spielberg's 2001 blockbuster *AI: Artificial Intelligence*. It's also one of the spookiest.

It's set in a chaotic future world in which climate change has done a lot of damage, and people have come to depend on Robots. They are called 'Mecha' – in contrast to 'Orga'. 'Mecha' are mechanical. 'Orga' – people – are organic. The reason it is so spooky is that Mecha have become a lot like Orga.

The story is complicated and I won't tell it all here. But at its core is the fact that a human child, Martin, is very ill, and his parents decide to get a Robot child to replace him. The Robot child is named David, and is programmed to 'love' them. But then David gets better, so they suddenly have two children. And then – surprise, surprise – things get difficult.

Martin's mother resists pressure to take David back to the Robot manufacturing company, where he will be 'killed', and instead takes him to the forest – and leaves him there. In the forest he joins other escaped and abandoned Mecha who live there free. But they have predators – humans who hunt them down.

The most dramatic scene in the film is of a wild country fair. Broken and discarded and runaway Mecha have been captured from the forest, and at the fair they are destroyed for fun in front of the crowd. David gets caught in one of these sweeps, and ... well, you will need to see the film to find out what happens next!

BE RIGHT BACK

Did you see *Be Right Back*? You should.

First broadcast in 2013 on Channel Four, it's part of the Black Mirror sci-fi series that has now been taken over by Netflix and given a global audience. The whole series is one of the best at explaining what our new technologies mean for us humans.

Be Right Back tells a gripping story, but like most gripping, disturbing tales it starts off with something very ordinary. A young couple, Martha and Ash, are moving into a house in the countryside. After they unload their stuff, Ash heads off to return the van they hired and pick up their car. He never arrives. Martha is frantically making phone calls when the tell-tale blue lights of the police car flash through the window.

She's grief-stricken, and her grief goes to a new level when she discovers she's pregnant.

Meanwhile, her friend Sara wants to help. She tells her about a new online service that will enable her to chat with her dead partner. Martha is horrified by the idea, but Sara signs her up . . . and soon she is chatting away. First online. Then on the phone.

The key to the technology is that everything Ash has ever posted online is scooped up by the computer running 'Ash' – the programme – and is used to give the kind of answers to Martha that Ash would likely have given.

Then we shift to another stage. 'Ash' tells Martha that, for an extra charge of course, she can go beyond chatting. And then a large box arrives, containing 'Ash' the Robot. After she has put 'him' in the bath and sprinkled some magic chemical dust in with 'him', Martha discovers she has a walking, talking version of her lost love. 'He' has limitations (he can only pretend to eat or sleep, for example), but all the time 'he' is learning better to imitate the man who died when the van crashed. I won't tell you how it ends.

WESTWORLD

This spooky series screened in the UK in 2016 and available from cable channel HBO was based on a movie from 1973. The original film was both written and directed by Michael Crichton (who also wrote *Jurassic Park*), and became something of a cult classic. It's about a theme park where you can go and stay on holiday. The ads proclaim: *'Boy, have we got a vacation for you!'*

There are actually three options at the park. One is *Medieval World*. Another *Ancient World*. And the third: *Westworld*. We get to glimpse each of them, but the main action takes place in *Westworld* – and this is the basis for the TV series follow-up. It's the old Wild West, complete with gun-slingers – and a brothel. Actor Yul Brinner stars as a gun-slinger, and things get interesting when something goes wrong with the programming and the humanoids get out of control.

The movie raises fascinating issues about how humans and humanoids should relate to each other. On the surface they all look alike. But if you are a tourist you can shoot the cowboys (there's a safety device on the guns to make sure you don't shoot any of your fellow tourists by accident). And you can use the brothel.

In the TV series the story gets more complicated but the principle remains exactly the same. Humanoid Robots mingle with tourists, both look exactly alike, and it's a rather mind-bending experience to follow the story and build some sort of connection with these two sets of characters, who look alike but are fundamentally different.

ROBOTS AHEAD!

If we want to understand what's going on, there are two things we need to do to be prepared.

One is to look around us at developments in Robot technology, and see how it is already being used and what kind of uses are not far off. In the chapters that follow we will be doing just that.

But the other is to look further ahead – which is why we have science fiction!

While plenty of sci-fi is completely out-of-this-world and the 'science' may seem ridiculous, plenty of it is not. It ties the imagination of its authors to technologies that are already being developed. That's one reason why watching and reading sci-fi can help us think through what may lie ahead.

And don't just read and watch for entertainment, although it's fine to enjoy the show. Think about what you see, talk about it, and do some online research to find out how realistic the scenarios are. Because some of them are getting too close for comfort.

QUESTION TO THINK ABOUT

How is that we can have had so much exposure to Robots in the media and yet so little discussion of their implications in the Church?

CHAPTER TWO

Aleksandra Suzi / Shutterstock.co

NOT ALL ROBOTS LOOK LIKE ROBOTS!

'With regard to robots, in the early days of robots people said, "Oh, let's build a robot" and what's the first thought? You make a robot look like a human and do human things. That's so 1950s. We are so past that.'
Neil deGrasse Tyson, Astronomer

Robot is a made-up word. It was made up in 1920 by a Czech writer named Karel Čapek. He was writing a play, and ended up with the name *Rossum's Universal Robots*. In the Czech language *robota* means something like 'slave labour'. It's an exciting story; we look at it in a bit more detail later. His factory turns out thousands and thousands of these 'slave' creatures, and *Rossum's Universal Robots* do pretty much all the work that there is to do.

When someone says Robot we all have in our minds the idea of a clunky human-looking machine, and there are certainly plenty of clunky human-looking machines being manufactured by inventors and labs all over the world. There are also animal-looking Robot machines – with four legs they are more stable. Walking has actually proved one of the most difficult challenges for Robot makers. We rarely think about it, but when we walk around the house our bodies are doing pretty complex things to keep us upright.

But it's important to realise that the Robot revolution isn't just about clunky machines that look a bit like people, or 'humanoids' as Robots like that are called. We've just seen that one of the most useful of the Robots already around – probably the one we use most frequently – is the disembodied voice that tells us where to go when we're driving. We may sometimes imagine a woman surrounded by maps working out the best route for us, but of course 'she' is just another programme.

The satnav system depends on GPS – the Global Positioning System (run by the US government) that co-ordinates a series of satellites up in the sky and gives us our bearings here on earth. If you are an Apple fan, you call that disembodied voice that gives you directions, and a lot more info if you ask, Siri. Android users call 'her' by other names. The point is: Robots don't all look like Robots. In fact not many of them do. The most common Robot in everyday use is the Roomba, the Robot vacuum cleaner that hoovers your carpet all on its own. It looks like a combination of a small flying saucer and a circular electric fire.

If you prefer, you can use a wider term like 'machine intelligence' or 'intelligent machines' to refer to these Robots that don't look like Robots, or may not have physical form at all. The key thing is that they are all applications of 'Artificial Intelligence' – they are fancy computers designed and trained for the specific tasks we ask them to perform. But I think it's simpler just to call them all Robots.

I've been making a list, and these Robots actually come in at least six separate varieties. They are all examples of machine intelligence designed for some specific task, or a collection of tasks. It's quite important that we understand that they are all basically the same thing. Here's my list.

ROBOT-LOOKING ROBOTS

First, there are Robots that look like Robots that look like people. At the moment, the best-designed and cleverest of them only look like people in a clunky, mechanical way.

We call these Robots 'humanoids'. Not many Robots are actually like this, though it's likely that many more will be in the future. It's mainly in Japan that inventors and investors have been focused on Robot-looking Robots (human-looking Robots). They have even put effort into making their faces respond like human faces do. Not easy – there are hundreds of muscles in your face. So all the time they are getting more humanoid. Elsewhere in the world the focus has been much more on what Robots can do than how they look.

Of course, as we have already seen, there are plenty of films about humanoids, from *Bicentennial Man*, with the irrepressible Robin Williams, to Stephen Spielberg's *AI: Artificial Intelligence*.

ROBOTS THAT JUST LOOK LIKE MACHINES

The second group are what we might just call robotic machines. They aren't designed to look like anything other than they are – just super-intelligent machines. We are familiar with all kinds of machines. From the time of the Industrial Revolution onwards machines have been central to much of our industry and, more recently, to our personal lives – cars from the early twentieth century – washing machines, vacuum cleaners and other labour-saving devices from nearer the mid-century. These machines are clever, but they are mechanical and at the end of the day pretty simple. The move from these basic mechanical machines to 'intelligent' Robot-type machines has been dramatic.

The latter half of the twentieth century has seen the installation of what are known as Industrial Robots in thousands of factories all over the world. They have often taken over people's jobs on production lines. They look like large tools, which is what they are, and – like the people who used to do the jobs – they are generally built and programmed to do only one or two things, like dropping an engine into a car chassis, or bolting on the doors.

Alongside Industrial Robots, we are now entering the age of the Domestic Robot. As mentioned the first one to arrive in our homes was the Roomba, the Robot vacuum cleaner that has now been around for a decade. It finds its way around your living room floor, avoiding obstacles and getting behind sofa legs and other items of furniture to keep your carpet clean.

Another Robot machine is your car! Already. You may not be aware of how 'intelligent' your car is, especially if it's a fairly new model. Long before the future shift to self-driving cars, our human-driven cars are packed with computer code. It's why they are far more efficient and reliable than cars used to be, but it's also why we can't easily service them ourselves and garages plug in their computers to diagnose the problem. Back in the 1960s, we used to push-start cars that had dead batteries, and wrestle with chokes, and with older models you could still crank-start them. Things have changed a lot. You can't do that now, though it would be pretty unusual if you needed to.

You may not be aware of how 'intelligent' your car is, especially if it's a fairly new model.

But maybe the prize should perhaps go to the humble cash machine, the ATM (automated teller machine), which is where most of us now do our banking. The high-end ones can now receive deposited cash and cheques as well as offer us statements and other account information. But for more than 30 years we've been used to getting our money out of holes in the wall.

A ROOMBA ROBOT VACUUM CLEANER

The point about these 'Robotic Machines' is that they each use Artificial Intelligence for just one purpose – running the machine, whether it's counting out cash, helping your car run more smoothly, or cleaning the floor. There is no effort to make them look like people, even though various of the things they do have displaced the jobs of the people who used to do them.

ROBOT TOYS, OR ARE THEY PETS?

A third kind of Robot, and one we shall be seeing a lot more of, are toys, though they already include toys for grown-ups which may better be seen as Robot pets. We recently saw the launch of a new Barbie doll that can chat with your children (more on her later). And there are experiments going on to see whether both children and old folk like having Robot pets around, to keep them company and perhaps monitor their health and their behaviour. Perhaps we can call them petoids. This will be a very big growth area in the future.

INVISIBLE ASSISTANTS

The fourth kind of Robot is the 'assistant' you never see but which responds like a human to voice and typed commands to bring you information and guidance. Our first experience of these Robot assistants consisted of automated voices on the end of telephones. Back 20 years ago when these first became popular they annoyed just about everyone who used them. But the technology has improved and we have generally got used to them. People often actually find them more convenient than having to deal with 'real' people if we want to pay bills or change our address or check on a delivery. And of course there is usually no waiting time. In the past few years there has been a huge improvement in the ability of Machine Intelligence to understand our speech.

Of course, Robot assistants now go well beyond telephone answering services. The satnav/GPS service for many of us is just the normal way to get directions. Using Siri and Cortana, Google and other assistants on our mobiles is still a bit hit-and-miss, but they're getting better all the time. There's a steady improvement in these technologies (perhaps ten per cent a year?), and whether we like it or not we shall soon be chatting to unseen assistants for all kinds of reasons.

PLAIN OLD ALGORITHMS

The fifth category is a lot less personal – much more the application of Robot intelligence in a totally machine-like way. The term that the experts use for the way that machines make decisions – the way they are programmed to, of course, by real humans – is 'algorithms'. It's an odd word but useful to know, as it refers to what seem to be the human-like abilities of all these machines.

So if you want to book a ticket on a bus or plane you will find a website that offers them. Then plug in your details. Then wait for it to search around – there may be several companies offering tickets that fit. Then once you make your choice you need to pay, and 'it' will connect you with your bank or credit card company. And all of this, if you like, while you are lying in bed or watching TV or Skyping with your Mum.

Of course, algorithms run everything. It was a programmer writing code who told your Roomba what to do when its sensors tell it that a sofa leg, or a sleeping dog, is in the way. A programmer has told the taped human voice (or cleverly humanized mechanical voice) which phrase to use as you approach a motorway and need to know which lane to use, or if you have missed your road and need to be turned around.

So the fifth kind of Robot may seem the most abstract – just the algorithm! No need to pretend to be like a person – just get the job done, and it's often a job that a person used to do. So when you go online to book a travel ticket, or pay a bill from your bank account, or 'like' something on Facebook, there are Robot algorithms in place, written by clever programmers, to decide what to do.

THE 'INTERNET OF THINGS'

The sixth and final kind of Robot is also totally impersonal, but it's becoming more and more important. It also has a very odd name: the Internet of Things. It basically consists of sensors that connect things you do in the real world with the internet – and helps you to control what's going on. The simplest examples are already in many people's homes, systems that control their central heating and other features and that can be operated from a distance – over the internet. These joined-up 'things' are meant to help us save money and time by giving us more control – whether of central heating systems, or fridges, or cars.

One way to personalise this a bit is to think of your home as a 'smart house'. That's the term being used by companies who want to help you have better control over all the systems that you have at home. So you can monitor the heat setting, check that the plumbing is OK (maybe you are away from home and there's a sudden freeze!), and turn on the oven so the roast will be ready for dinner.

Some of these ideas seem a little silly – like having the fridge tell the supermarket when you need more milk, or maybe just text you when you're driving home to remind you to pick some up. But it's surprising how fast we adapt to these new conveniences.

The point is: your whole house is turning into a kind of Robot, talking to itself, talking to you, communicating with your energy and phone companies. On a wider scale, 'smart cities' are all the rage. Whole city-wide Robot networks controlling street lights, monitoring pollution, watching out for crime – and maybe dropping you a hint if you're using more power than other houses in the areas. Add them all together, and you have a 'smart world'. One whole worldwide Robot.

The basic point is that these are all Robots, or, if it is more helpful to put it like this, they are all examples of ways in which Intelligent Machines are becoming part of our everyday lives. Paying your bills online may seem very different from buying a smart pet to keep your children occupied or hiring a Robot-looking 'humanoid' to put granny to bed, but the same technology is at work. Our million-times-more-powerful-than-NASA-in-1969 smart mobiles are the result of almost unbelievable technological progress that is now smart enough to take on a huge range of tasks that were previously things only smart people could do – or nobody could do.

The big question is: what will it mean as these machines become more and more 'intelligent,' and more and more and more like us? Because they will, and it won't be long. This is far and away the most challenging question of the twenty-first century.

Remember, God gave us dominion – he put us in charge. We've been busy making tools. At first they helped us farm – the adze, the wheel, the plough. Then they helped us make things – John Kay's Flying Shuttle in 1733, Hargreave's Spinning Jenny in 1764. Then they helped us get around – the Steam Engine, perfected by James Watt in 1764, and the Aeroplane. Now? They are helping us do pretty much everything, and they are starting to do some things a lot better than we can.

> *Remember, God gave us dominion – He put us in charge.*

So the challenge is: after *making* them, what do we *make* of them? The Them and Us challenge, we could call it. We shall need to work out what to do with these increasingly intelligent Robots. Some of them already chat with us, and many more will. Some will look like us, and work for us and with us. This will make heavy demands on what we think and what we believe. And how we think about all this will shape how we live, in pretty much every aspect of our lives.

QUESTIONS TO THINK ABOUT

Does it help to lump together all these different Artificial Intelligence systems – from cash machines to Siri to humanoids? Are they really all examples of the same thing? Are human-looking Robots really different?

CHAPTER THREE

WHAT MAKES US SPECIAL?

'So God created mankind in His own image, in the image of God He created them; male and female he created them. God blessed them and said to them, "Be fruitful and increase in number; fill the earth and subdue it. Rule over the fish in the sea and the birds in the sky and over every living creature that moves on the ground.' Genesis 1:27, 28

Christians have a clear idea of what it means to be one of us – a member of our species. We believe that God made us, and that He made us in his image. But unpacking what Genesis means by *in His image* is not so simple. Plainly, we aren't God, so it can't mean we are exactly the same as He is. But we also aren't animals, or birds, or fish, or any of the other things that make up his creation. We find ourselves somewhere between what they are, and who He is.

Fortunately, there is further light shed in Scripture on what Genesis means by 'in His image'.

First, Psalm 8. The words are a kind of commentary on the creation story in Genesis, and its importance is reinforced by being quoted in Hebrews 2:7, 8.

'When I consider your heavens,
the work of your fingers,
the moon and the stars,
which you have set in place,
what is mankind that you are mindful of them,
human beings that you care for them?
You have made them a little lower than the angels
and crowned them with glory and honour.
You made them rulers over the works of your hands;
you put everything under their feet:
all flocks and herds,
and the animals of the wild,
the birds in the sky,
and the fish in the sea,
all that swim the paths of the seas.'
Psalm 8:3-8

'You made them a little lower than the angels;
you crowned them with glory and honour
and put everything under their feet.'
Hebrews 2:7,8

Our responsibility for the creation and our being made in God's image go hand in hand. Whatever else being made 'in His image' means, it enables us to serve as His stewards in managing the earth.

Christian thinkers have explored the meaning of our being in the 'image' of God in various ways. What is it that makes us humans unique? There are all kinds of dimensions to bearing the image of God – ways in which we are like Him. Let's summarise them under five headings.

First there's a focus on our brains – our ability to reason. But not just reason in the sense of argue. We can think things out, we can reflect, and this becomes the basis for everything else we can do.

Second, the human brain is also the source of our ability to make choices – we have free will. So we are moral creatures, not just calculators.

Third, we can also reach beyond ourselves to be creative in art and poetry and music. The package of human qualities that flows from our brains reflects what God himself is like.

Fourth, we are called to live on this Earth on God's behalf. We are here as His stewards, and His giving us dominion over the Earth is all part of our reflecting his rule over the universe.

We are creatures who have relationships – human relationships, personal relationships – just as God has a relationship with us.

Fifth and finally, we are creatures who have relationships – human relationships, personal relationships – just as God has a relationship with us.

While some thinkers have emphasised one of these more than the others, they fit well together, and together they fill out the picture of us humans as being made like God – but in the limited confines of time and space. God has given us brains that can reason, make moral choices, and also be creative. He has placed us here to rule the earth on his behalf. And He has made us creatures who thrive in relationships, with each other and with him. These are all strands in our being made 'in His image' – space-time models of God himself.

Yet the most dramatic claim made by Christians is not that we humans are made in God's image, it's that having made us in His image God then took the extraordinary step of coming down to join us.

That's what theologians calls the Incarnation of Jesus Christ – which literally means 'being put into flesh' or 'into bodily form.' While we tend to focus on the wonder and joy of Christmas, behind the story of the Angel Gabriel appearing to Mary, the journey to Bethlehem, the arrival of the shepherds and later the wise men – behind all this is the greatest claim that anyone has ever made. That God, having made the cosmos, decided to visit our planet and live a life like the lives we live. In the process he gave an almost unbelievably special quality to our human existence. Aside from the fact that we sin and Jesus never did, He led a life just like ours.

Well, not 'just like'. It was a unique life – of teaching, example, and miracles. It seemed to end on the Cross, where He paid the price for sin. His resurrection brought his first followers face to face with the Jesus they had known and loved, and yet He was also different. And soon after that He 'ascended' into heaven – disappearing before their eyes, as He returned to the right hand of God the Father.

Now just to say that is to admit that it is all very mysterious, and the greatest minds in the history of the Church have wrestled with what it really means. How exactly had his body changed when he arose from the dead? It puzzles us that he could pass through doors and yet also eat. And plainly, when he 'ascended' He was doing something symbolic, as heaven is not floating around in the stratosphere!

Something we often neglect is the fact that the 'Incarnation' – the Son of God's bodily life in Palestine, as Mary's son also – has not come to an end. In theological terms, the Ascension does not mark the end of the Incarnation any more than the Crucifixion and Resurrection did. The 'human' life that Jesus started has no end. Charles Wesley, the great Methodist hymn-writer, put it like this:

'Of our flesh and of our bone,

Jesus is our brother now.'

The amazing truth of the Incarnation underlines the special-ness of members of the human species in a manner that goes beyond our understanding. Humans are so special that the Son of God decided to take human nature back with Him to heaven. How He can be both human and divine at the same time we don't know. But it's clear from the Bible that he is. Whatever God making us 'in His image' means, it tells us that we were made enough like Him that he could then step into our human shoes.

So: there's something uniquely special about our human life, and that's why these rapid advances in technology raise such big questions.

Machines are great as tools, and Robots are the ultimate machines. But they are tools to enable us to be God's image-bearers here on this planet. How can they help us, and not hinder us, in these five 'image' roles God has given us? As thinking, creative beings. As those whom He has called to run things down here for Him. And as women and men designed to have relationships – with each other and with him? These are questions to have in our minds as we read what comes next. We'll come back to them at the end.

BUT WHAT ARE WE HERE FOR?

That question can be answered on two different levels. On one of them, plainly, we are here to praise and worship God, both with our voices and with our lives. But there's another level. What are we supposed to be doing with those lives, aside from generally leading them in a way that will please Him?

The usual answer is a simple one: working! Not all in the full-time workforce, of course. Some are called to be homemakers, and certainly for part of the time many of us – usually mothers but also fathers – take time out of the workforce to focus on bringing up our children.

And then, at least in the West and in recent times, we get to retire. People are living a lot longer now than they used to, so there's a reasonable prospect of ten or 20 years of life after full-time work has stopped.

Christians have taken work seriously not simply as a necessary way of earning a living, but as a way of serving God. There are two sides to it.

Christians have taken work seriously not simply as a necessary way of earning a living, but as a way of serving God.

First, it's something God commands us to do. After God had made Adam, back before the fall, he was placed in the Garden of Eden to look after it:

> 'The Lord God took the man and put him in the Garden of Eden to work it and take care of it.' **Genesis 2:15**
>
> And after the fall:
>
> 'To Adam he said, "Because you listened to your wife and ate fruit from the tree about which I commanded you, You must not eat from it,
> Cursed is the ground because of you;
> through painful toil you will eat food from it
> all the days of your life.
> It will produce thorns and thistles for you,
> and you will eat the plants of the field.
> By the sweat of your brow
> you will eat your food
> until you return to the ground,
> since from it you were taken;
> for dust you are
> and to dust you will return.'"
> **Genesis 3:17-19**

In other words, Adam had been given work to do in the first place, but because of sin it is going to become harder and a lot less pleasant. So what will it mean if Robots start doing our work for us?

Back around 1930 two of the most influential thinkers of the twentieth century were aware that Robots were on their way – and would be taking work from humans.

John Maynard Keynes, the great British economist (and without doubt the most influential economist of the past two hundred years), wrote about a future in which we would perhaps only have 15 hours of work a week each. And he looked ahead to the birth of what he called (no doubt with a smile on his face as he wrote it) 'a new leisured class'. In other words, ordinary folk who didn't have to work. We'll come back to what he had to say when we think about the possibility of a World without Work.

The other man was less well-known but extremely influential. His name was Norbert Wiener, and he was a maths professor at the Massachusetts Institute of Technology. Aside from being a brilliant mathematician – he was awarded his undergraduate degree at the age of 14! – he is known as the 'father of cybernetics'. Cybernetics is basically the science behind technologies that control things – like Robots. Professor Wiener looked ahead to a time when technologies would take work away from us, and perceptively – and provocatively – described what was coming as a 'slave economy'. In other words, Robots are slaves. And when slaves are competing against free people in the economy, slaves will always be cheaper and get the jobs.

So there are two questions that we need to face here. First, how will we serve God if we don't have jobs? And second, and more generally, what on earth are we going to do with ourselves? We come back to these questions in Chapter 12, after taking a closer look at what Robots are up to.

QUESTION TO THINK ABOUT

Think about people you know and love. What is there about them that a machine couldn't copy?

CHAPTER FOUR

A QUICK LOOK AT THE BIG PICTURE:
WHERE'S TECHNOLOGY GOING?

'The concept of the robot encapsulates both aspects of technology. On one hand it's cool, it's fun, it's healthy, it's sexy, it's stylish. On the other hand it's terrifying, it's alienating, it's addictive, and it's scary. That has been the subject of much science-fiction literature.'
Thomas Bangalter, French musician

Technology seems to be exploding all around us, and it isn't all about Robots.

For example, Nanotechnology – which basically means tiny technology (*nanos* is actually the Greek word for a dwarf) – has been all the rage for the past 20 years. It means engineers can do things with matter in a much more basic way than used to be possible. Already there have been all kinds of uses, including some that have probably benefited you directly – like more effective suntan lotion using the chemical titanium dioxide. (Check the small print next time you buy some and see.)

Biology is also being transformed by bio*technology*. New kinds of drugs have become possible. The idea of 'personalised medicine' – with medicines designed just for you – is finally becoming a practical proposition. Maybe the most exciting (and alarming) new field is that of 'synthetic biology', where scientists basically build living things out of a kit made from parts of others.

It's alarming partly because Frankenstein-type creations might become possible, but also because the equipment involved is relatively basic – you don't need to spend billions of pounds on a laboratory. Terrorists and other 'bad actors' might be able to get in on the act.

At the moment the most controversial technique in genetics is called CRISPR (Clustered Regularly Interspaced Short Palindromic Repeats). It's a very recent discovery that has suddenly meant we can 'edit' the genes that programme plants and animals and people relatively easily. Scientists are already working on engineered mosquitoes to try and stamp out tropical diseases.

Much of this activity in various sciences is being powered by Information Technology (IT). As you know when you try and sort out your Christmas card list, once you put information into digital form it becomes a lot easier to manage it. That same principle has revolutionised biology and many other areas of science. And as we discuss elsewhere in this book – and as you know from your smartphone and how much smarter it is than the one you used five years ago – IT is moving forward at an incredible pace.

Cognitive Science, or Neuroscience, the study of the brain, is also a bubbling cauldron of new discoveries and fresh theories. Of course, one reason it is so important is that some of the scientists working on building Robots want to copy how humans think. Those who are more 'out there' – planning for the day when, they reckon, the machines will become as smart as we are – are also fascinated by ways in which we can join up human brains and machines. As sci-fi fans know, the term for that is a 'cyborg' – an abbreviation of 'cybernetic organism'. Cybernetics, remember, is a fancy term for using computers to control things.

Of course, most scientists and technologists working on Robots don't much care about how the human brain works. What interests them is getting the job done. Self-driving cars and cash machines work very differently from human drivers and bank tellers, but they navigate the road and count out the cash anyway. As famous computer scientist Edsger Dijkstra once said: 'The question of whether machines can think ... is about as relevant as the question of whether submarines can swim.'

The question of whether machines can think ... is about as relevant as the question of whether submarines can swim.

If you look at each of these areas of science and technology in turn, you realise quite how revolutionary things are getting.

The US government's National Science Foundation organised a series of conferences back in the early 2000s to explore how all these pieces might fit together, and came up with the acronym NBIC for Nanotechnology, Biotechnology, Information Technology, and Cognitive Science.

BILL JOY: THE FUTURE DOESN'T NEED US

Meanwhile, one of the gurus of the world of modern technology, Bill Joy, a founder of computer giant Sun Microsystems, astonished many of his colleagues with an article in *Wired* magazine – the monthly of the tech community – that threw all this into question. Published in April of 2000, it had the stunning title: 'Why the Future doesn't need us'.

Bill Joy's argument was a bold one. He laid out his concern that, one way or another, humans might not have much of a future. There were two serious possibilities.

One was that we would destroy ourselves, whether through a ghastly accident or perhaps the efforts of bad people. Or – and this is where his warning rings true in our discussion of the future of Robots – we would make a non-human future by making Robots who are smarter than we are.

The article is summed up in its opening paragraph: 'Our most powerful twenty-first century technologies – robotics, genetic engineering, and nanotech – are threatening to make humans an endangered species.'

Joy explains that he really hadn't thought much about these issues until a couple of years before, when he happened to meet futurist and inventor Ray Kurzweil. Kurzweil's focus on the way in which Robot technologies are moving ahead faster and faster caught his attention. (We discuss Kurzweil's key idea of the 'singularity' in a few pages' time.) He began to reflect on other tech developments with a more critical eye. For example, one of the possibilities that futurists hold out for Nanotechnology is making tiny machines that will build things for us using raw materials. But things could go wrong, and if they do then these tiny machine bugs could eat their way through absolutely everything. This has been termed the 'Grey Goo' scenario – the idea is that they could turn the whole world into goo. Grey Goo became famous in the UK when Prince Charles talked about it some years back.

Bill Joy's point is that things can go wrong, badly wrong. We certainly don't need to be 'anti-technology' to raise difficult questions about what could lie ahead. Joy has now moved on from helping run his company to becoming a venture capitalist, and investing in other technology efforts. His words of warning, coming from a top tech insider, are words we should listen to. (You may like to read his article. It's not all that difficult, and you can get it free online – wired.com/2000/04/joy-2/)

It's worth keeping an eye on the Big Picture.

KEVIN WARWICK: THE 'WORLD'S FIRST CYBORG'

Just two months before Bill Joy's warning, British professor Kevin Warwick was making a big name for himself – in the same place, the front over of *Wired* magazine. He's been called 'the world's first cyborg' – and his experiments on himself certainly show that humans and computers can be connected.

The interview in *Wired* begins by being even more provocative.

'I was born human. But this was an accident of fate – a condition merely of time and place. I believe it's something we have the power to change. I will tell you why...

In August 1998, a silicon chip was implanted in my arm, allowing a computer to monitor me as I moved through the halls and offices of the Department of Cybernetics at the University of Reading, just west of London, where I've been a professor since 1988.'

But that was just the first step in Project Cyborg. Next:

'On the 14th of March 2002 a one hundred electrode array was surgically implanted into the median nerve fibres of the left arm of Professor Kevin Warwick. The operation was carried out at Radcliffe Infirmary, Oxford, by a medical team headed by the neurosurgeons Amjad Shad and Peter Teddy. The procedure, which took a little over two hours, involved inserting a guiding tube into a two inch incision made below the elbow joint, inserting the microelectrode array into this tube and firing it into the median nerve fibres above the wrist.'

His wife Irena also got an implant, and as a result:

'As Kevin Warwick gently squeezed his hand into a fist one day in 2002, a robotic hand came to life 3,400 miles away and mimicked the gesture. The University of Reading cybernetics professor had successfully wired the nerves of his forearm to a computer in New York City's Columbia University and networked them to a robotic system back in his Reading, England, lab. "My body was effectively extended over the Internet", Warwick says.'

The linking-up of humans and machines to create a kind of hybrid is another part of the Big Picture.[1]

RAY KURZWEIL: PROPHET OF THE 'SINGULARITY'

The man who sparked Bill Joy's interest and anxiety was Ray Kurzweil, a brilliant inventor and futurist who is now working with Google.

He has some strange ideas, but before we write him off as a nut, we should listen to Microsoft founder Bill Gates. Gates has described Kurzweil as 'the best person I know at predicting the future of Artificial Intelligence'. Ray Kurzweil has made it his life's work to forecast the coming of what he calls 'the Singularity.' This is the point when (if it ever does happen, and of course we don't know) Robot intelligence gets smarter than humans. Kurzweil has been predicting this for a long time, and he's written a huge book about it called *The Singularity is Near*. His latest prediction is that it will come about in the year 2045.

The word 'singularity' can actually mean various things, but Kurzweil took it with this special meaning from a short story by computer science professor and sci-fi writer Vernor Vinge. Kurzweil's book is full of scary and exciting and rather breathless statements like these:

'Although the Singularity has many faces, its most important implication is this: our technology will match and then vastly exceed the refinement and suppleness of what we regard as the best of human traits.

Our sole responsibility is to produce something smarter than we are; any problems beyond that are not ours to solve.

Our human intelligence is based on computational processes that we are learning to understand. We will ultimately multiply our intellectual powers by applying and extending the methods of human intelligence using the vastly greater capacity of nonbiological computation.'

It's fair to say that Ray Kurzweil isn't just a brilliant engineer and inventor, but he's also an optimist. He's an optimist about the rate at which technology will be able to get us where he believes we are going. People have made such predictions in the past, and things simply haven't worked out. But, more important, he's also an optimist about humans and the Robot machines we are making and how all this will turn out. He's excited about what will happen when these machines, built by us and with our values built into them, take over and solve our problems.

At the conference in California some years back where I got to meet with Kurzweil I was also able to meet the sci-fi writer Vernor Vinge. Vinge had been speaking at the conference, and one of the things he said to the audience has stayed with me ever since. It struck me that he was more of a realist. He's an enthusiast for our technological future, but he may be less naïve in his optimism than Kurzweil. Here are his sobering words: 'The longer we keep our hands on the tiller, the better'.

QUESTION TO THINK ABOUT

How is it that top people in the technology world can come up with such different views about what is going to happen, and whether we should be worried?

CHAPTER FIVE

A QUICK LOOK AT THE BIG PICTURE: OTHER INTELLIGENT BEINGS?

'Are Whales Smarter Than We Are?' ***Scientific American***

One reason why the Bible's emphasis on our being made 'in His image' is so important is that, obviously, it separates us from smart machines – something we will be looking at in detail later in this book. But there is another, equally important, reason: it separates us from other beings.

ANGELS

It may seem that humans are the only smart creatures around, but that's not the case. For one thing, there are angels. The Bible does not say a great deal about them, and we need to be careful not to speculate too much. But we do know God made other intelligent beings, though He did not make them of flesh and blood.

ANIMALS AND BIRDS

And what about the animals? Many people assume that the other creatures God made are basically stupid, little more than things with legs. We use the phrase 'dumb beasts'. Yet as farmers know, and anyone who has pets, animals have plenty of intelligence. Some are smarter than others. Goats are a lot smarter than sheep. Dogs and cats have intelligence of very different kinds, and while cats can be very affectionate (when they feel like it!) dogs can form bonds with humans that make them very much part of our families, and friends for life.

In fact, recent scientific discoveries have provided astonishing new details of how intelligent some animals and birds really are. Dog owners have long known that man's best friend can be trained to respond to commands – and also needs no training to slink away and look guilty when you arrive home if he or she has done something bad!

But some of the science is amazing. One researcher trained his dog to recognise one thousand different toys – and bring the one he asked for almost every time. Another study recently showed that when dogs seem to be listening to you they are not – as we used to believe – just responding to your tone of voice. They are picking up something of what the words mean, and it makes a difference if you say in the same tone of voice 'good dog, wonderful, I love you' and 'stupid beast, we need to get rid of you'. As dog owners know, some breeds show a lot more intelligence than others, and there can be big differences between one dog and another even in the same litter – as with humans! The point is that there's a lot more going on in a doggy brain than we used to believe.

Some of the most surprising research involves birds. Just like we speak about 'dumb beasts' we use the term 'bird-witted'. But birds – some birds, especially – certainly have their wits about them. We've always known that parrots were smart – they aren't just mimics, 'parroting' things people say, but can learn the names of objects and even engage in basic conversations.

The most celebrates case was that of 'Alex', whose thirty-year life was studied by American scientist Dr. Irene Pappenberg. She found that he was as intelligent in some respects a five-year-old child, and had the emotional life we find in a human two-year-old. He was so famous that when he died at the age of 30, *The Economist* published his obituary! His death was a sad surprise – parrots can live till they are 100.

But it's not just parrots. Other bird species have been discovered to call to each other by the equivalent of first and last (family) names. And 'corvids' – the bird family that includes crows, ravens, and magpies – have astonished researchers by their abilities. For example, once they have seen your face they can remember you years later. If they want to crack a nut open, they may leave it on a zebra crossing – so a car will run over it for them, and they can then fly down when people are crossing and it will be safe to retrieve!

The intelligence of the great apes is better known, and the fact that they look like us makes it simpler for us to expect that sometimes they will behave like we do. Of course, they are also wild animals and can be dangerous. But researchers like Jane Goodall have spent their lives studying them and shared what are sometimes bizarrely 'human' stories about their lives. In a different way the same is true of elephants, which makes their being killed for their tusks a special tragedy. They live long

in family groups, they grieve when one of their number dies, and there are remarkable stories of their relations with humans who have grown close to them.

The final group of 'intelligent' non-humans are the family of species known as cetaceans. They include dolphins, but also various species of whales some of which seem to be much smarter than others. It's no surprise that researchers have found it very difficult to work with these animals. Dolphins are the exception, and many people believe them to be smarter than chimpanzees.

Some researchers believe the sperm whale – Moby Dick was a sperm whale – may actually be the smartest of all the animals, aided by possessing a brain five times as large as ours. Indeed, scientists have sometimes wondered if the sperm whale could in reality be more intelligent than human beings. But these comparisons are complicated and depend on what we mean by intelligence, and how well we understand what it means to be something very different from us. Try imagining for a few minutes that you are a smart sperm whale, and you will get the point.

The octopus is another good example. Bizarrely, it has brain cells in its tentacles, and we have no idea what that means. Octopuses are also good at remembering people they have met before. One squirted a researcher as she walked into the building years after it had last met (and not liked) her!

Octopuses are also good at remembering people they have met before. One squirted a researcher as she walked into the building years after it had last met (and not liked) her!

EXTRA-TERRESTRIALS?

It's a big jump from angels and whales to the inhabitants of other planets, but if we want to make sense of 'intelligent' Robots it may be helpful to consider all the intelligent non-humans who may be around. Of course, we've no means of knowing if there are intelligent beings elsewhere in the universe. But it's an age-old question that has rather suddenly acquired a lot of interest on the part of scientists – mainly because we're becoming a lot better at interpreting the information we get from far out in our galaxy, the Milky Way.

Since the start of space exploration in the mid-twentieth century we have sent probes to all the planets in the Solar System, and to other bodies such as their moons and asteroids and comets. But the stars that might be shining down on planets like the earth, planets that could harbour intelligent life, are far, far distant. Recent research has suggested that there are in fact many such 'earth-like' planets – close enough to their stars to be warm, but not too close!

While the Bible's focus is on our planet and our species, it nowhere denies that there could be life elsewhere.

For a long time many people believed that the possibility of intelligent life was low. One reason was that if life just emerges there would be civilisations far older than ours on some such planets – and they would have been advanced enough to come and say hello. Since, so far as we know, they haven't, the argument goes, they can't be there. The great Italian physicist Enrico Fermi – famous for having built the world's first nuclear reactor – pointed this out, and it has since become known as the Fermi Paradox. Billions of stars, likely millions of planets, some far older than ours, and no visitors. 'Where are they?' he asked.

Now we are getting much more information about far-distant planets the arguments have begun to swirl. One recent book, issued with a Foreword by Astronomer Royal Lord Rees, came up with no less than 75 different actual answers to Fermi's question!

Of course, Christians may take the view that because the Bible says nothing about other intelligent creatures and civilisations then there are none. We are unique in the universe – a conclusion that the Fermi Paradox could be taken as supporting. On the other hand, while the Bible's focus is on our planet and our species, it nowhere denies that there could be life elsewhere – as C.S. Lewis explored in his science-fiction

Space Trilogy novels. What Christians should make of an extra-terrestrial intelligent species is another puzzle to add to the puzzles we are discussing in this book.

But as we move on to think more deeply about Robots it is useful to remind ourselves that we know there are other intelligent life-forms in the universe. There are undoubtedly angels (as well as fallen angels, or demons). There are animals and birds that we now know to be far from 'dumb beasts' and 'bird-witted' even though they seem less smart than we are, and their smartness is certainly of a different kind. And – there may be other intelligences on other planets.

QUESTIONS TO THINK ABOUT

Does it worry us that God made other intelligent creatures? Whether they are parrots or sperm whales or maybe extra-terrestrials. How are we still special?

CHAPTER SIX

WHERE DID ROBOTS COME FROM?

'Let's start with the three fundamental Rules of Robotics... We have: one, a robot may not injure a human being, or, through inaction, allow a human being to come to harm. Two, a robot must obey the orders given it by human beings except where such orders would conflict with the First Law. Three, a robot must protect its own existence as long as such protection does not conflict with the First or Second Laws.' Isaac Asimov

Robots aren't new. While we are getting much better at making them, machines that do smart things have been around for a long time. Some of them were real, some of them were just imagined, and one of them puzzled some of the greatest minds in the world as they tried to understand how it worked.

People used to be call these machines 'automatons' – because they would do things automatically, or, literally, on their own. Then, as we have already seen, in 1920 Karel Čapek came up with a new word for them, and it quickly caught on.

'DELIVER US FROM THE ROBOTS'

In fact as mentioned before, Karel Čapek's new word in the Czech language, robota, means something like 'slave labour'. In his play, *Rossumovi Univerzalni Roboti* he had an eye to an English-language audience from the start, so he gave it an English subtitle: *Rossum's Universal Robots.* And we've been calling them Robots ever since.

His play tells the story of an inventor named Rossum who was a kind of Czech version of Dr. Frankenstein. Rossum was a mad scientist who dabbled in his lab until his experiments finally worked – and he found a way to make something that looked and acted very like a human. Unlike Dr. Frankenstein, he also had an eye to business. So he built a factory that could turn out thousands of them. Rossum's Robots were less mechanical than today's, and rather more like real people. But as they left the factory they were put to use as the slave labourers they were, hundreds of thousands of them.

Rossum's Universal Robots is an exciting story. All kinds of things are going on in the Robot world that Rossum's invention has brought about. There are well-meaning people trying to get 'human rights' for Robots, and other well-meaning people trying to get them to rise up in revolution. We hear about one revolt that was put down by humans. And we read about the problem that having great masses of 'slave labour' Robots creates for humans who want to have jobs.

And we even have what is probably the first Robot prayer ever written! One of the human beings, called Alquist, a key manager of the Rossum's Universal Robot factory, is having second thoughts. He's especially concerned about the way Robots have taken away people's jobs. This is what he prays: 'Oh Lord, I thank thee for having given me toil. Enlighten Domin [the Robot factory boss] and all those who are astray; destroy their work, and aid mankind to return to their labours ... deliver us from the Robots.'

THE MECHANICAL TURK

Way back in 1770, an Austrian aristocrat called Wolfgang von Kempelen decided to impress the Empress Maria Theresa of Austria by building a Robot that could play chess. For the next 84 years the extraordinary Robot chess-playing machine that was called the Mechanical Turk, or just the Turk, beat chess masters all over Europe, even Napoleon and Benjamin Franklin.

It was officially known as the automaton chess-player, but the Turk name soon caught on. It consisted of a box nearly three-feet square with, on top, the life-size model of the head and chest of a man with a black beard, grey eyes – and Turkish costume. On top, of course, there was also the chessboard. As the challenger played – sometimes on the board itself, though apparently when Napoleon played he used a separate board some distance away, the Turk would respond by moving his arm and hand and his piece. And the Turk almost always won. The first challenger was one of the Empress's courtiers, Count von Cobenzl. He was quickly beaten.

If the Turk threatened his opponent's queen, he would nod twice. If he threatened the king with check, three times.

Photo: Marcin Wichar, Creatrive commons Attribution 2.0 Generic

If the Turk threatened his opponent's queen, he would nod twice. If he threatened the king with check, three times. If his opponent made an illegal move, the Turk would move it back and make his own move. While he played, von Kempelen would move around the audience inviting people to challenge the automaton. Perhaps it is worked by magnetism? Then bring your magnets! And if you wanted to ask the Turk questions, you could. He would respond (in English, French, or German) by placing letters on a board. It must have been a lot of fun to watch.

Meanwhile, Kempelen was distracted – he was an ambitious inventor, and among other things was working on ways to replicate human speech. But in 1781 the Emperor Joseph II demanded that he put the Turk to work again. The Grand Duke Paul of Russia was to be visiting him, and the Emperor wanted to show off. So Kempelen brought out the Turk again, and played yet another game of what looked like magic. And for years to come, on and off, the Turk impressed audiences all over Europe. Then in the 1820s he crossed the Atlantic, and toured the United States. Sadly, we can't examine the miraculous machine ourselves. It perished in a fire in 1854.

The Turk was a con, from start to finish. One reason the story is so fascinating is that it fooled so many people for so long. And all the time the many chess players and other staff who were in on the trick kept the secret. All through the Turk's life of 84 years – and beyond! Back then, of course, they didn't have Facebook or Twitter, so secrets were certainly easier to keep.

Finally, the truth came out. In fact Kampelen, and the Turk's owners after him, recruited a chess master who was small – at least one was a woman – swear them to secrecy, and cram them inside the box. The cabinet was very cleverly constructed with cogs and other mechanical-looking elements, mostly just for show. The chess-player hidden inside had a sliding seat so he or she could flip between compartments if the owner was opening them up, to show the audience that there was no-one there. The pieces were moved by magnetism, and there was also a system to enable the chess-player to communicate with the person outside controlling the system. As con-jobs go, it was pretty smart.

Not until the late twentieth century would computers be able to play chess for real. And it wasn't until 1996 that a computer would really beat a chess champion. That was IBM's famous Deep Blue, their top-end machine, and the player it beat was Gary Kasparov, the famous Russian grandmaster. In fact, while Deep Blue won the first game of a six-game series, Kasparov then won three and two were drawn. So the human actually beat IBM's best in the series. It was the next year before the computer, which had been upgraded by the company, finally, but only just, managed to beat Kasparov, the reigning world chess champion, by the slimmest possible margin of 3.5 to 2.5. Kasparov then accused IBM of cheating, and demanded a re-match, but the company refused. Deep Blue was put in a museum.

It's worth noting that 20 years have now gone by since machine intelligence proved it was smarter than the best human being at what most people agree is the brainiest thing a human can do. A lot has happened since then, and, of course, it keeps happening faster.

JEOPARDY!

Robot game-players have since scored two more big victories against human intelligence. The best-known was in 2011 when another giant IBM computer, this time called Watson, named after the company's founder, not Sherlock Holmes' side-kick, beat the two top champions of the American TV game-show Jeopardy.

Ken Jennings, who had won 74 games in a row on the show, commented: 'I, for one, welcome our new computer overlords'.

While it might seem silly to compare winning a TV game show with beating the world's top chess player, it's actually been very difficult for computer engineers to come up with machines that can answer questions put to them in 'natural language' – which means, how you and I speak. So the Jeopardy win was seen as a big deal – a 'vindication for the academic field of artificial intelligence', wrote the *New York Times*.

GO

Even more important, in 2016 a computer programme called AlphaGo originally developed by Google was able to beat Lee Se-Dol, one of the very top players in modern times of the complicated Asian game called Go. Many people think Go is a bigger challenge than chess, and certainly a bigger challenge for a computer. Once again, the computer and the human played a series of games. AlphaGo won 4-1.

GORDON MOORE AND HIS LAW OF CRAMMING

While for hundreds of years people dreamed about high-intelligence Robots and, sometimes, pretended that they really existed, it's only very, very recently that the kind of machines we read about in *Rossum's Universal Robots* have become serious possibilities.

The reason lies in the same technology that has given us mobile phones. Back in the mid-1960s, a man called Gordon Moore made a statement that has become one of the keys to the future. He was writing about his business, which was building computer chips. (He was co-founder of Intel Corporation – the company now known to just about every computer user by those little 'Intel inside' stickers.

Gordon Moore's comment was that every year it has become possible to fit twice as many circuits onto a chip – the tiny silicon component that now may include millions of them, because we can make them smaller and smaller. In fact, the word he used was 'cram'. His comment that we can cram a whole lot more circuits onto our chips quickly became known as Moore's Law – the principle that every single year the chips keep doubling in power and computer technology gets far, far more potent.

What is remarkable is that 50 years later this 'law', while it may be slowing down a little, is still pretty much on target. The result of all that microscopic 'cramming' over all those years has been, well, astronomical! So, as we know, our mobile phones are around one million times more powerful than the computers that sent Neil Armstrong to the Moon in 1969. In fact, it's been calculated that, since the 1950s, our computing power has actually increased by a factor of *one trillion* – a 1 followed by 12 zeros. The point is: everything people used to dream about and imagine and fake in the world of Robots is either now true already, or is close to becoming true, or at least may be possible before long.

There is something very strange – and definitely scary – about this faster-and-faster process. It's what mathematicians call exponential change, or compounding. And while we can see it around us in the ways our phones keep getting smarter, it's still best illustrated by the old story of the Indian king and his chess-board.

Not, this time, the chess-board of the 'Turk' or grandmaster Gary Kasparov; any old chess-board, or draughts-board, will do.

The story goes that an Indian king used to like challenging visitors to a game of chess. To encourage one such visitor, the king said he could pick what he would like as a prize if he won the game. The wise old man said he didn't want much, but would like to win some rice – the quantity of rice you get when you put one grain of rice on the first square of the chess board and double it to two on the second, and then right across the board, doubling each time.

It seems a simple enough request – until you do some maths. Pull your calculator out, and follow along. Four grains on the third square, eight on the fourth,16 on the fifth. And then the numbers quickly get a lot larger. By the end of the first row you are up to 128. By the end of the second, 32,868. And then they get ridiculous. By the end of just the third row you are up to 8 million. And of course there are five more rows to go.

By the time you reach the final square, the maths has become unbelievable. It's 18,000,000,000,000,000,000 grains of rice. Apparently that comes to around 210 billion tons of rice, which is enough – someone suggests – to cover the whole of India with rice one metre deep... (I didn't work this out – it's all been calculated for us on the website singularitysymposium.com.)

NASA KENNEDY SPACE CENTER MUSEUM, APOLLO MISSION LAUNCH CONTROL

Linda Moon / Shutterstock.com

The point is: the pace of change being driven by digital technology is already almost unbelievably fast – but the prospect is that it will keep getting fast faster, if you can get your brain around that idea.

NEVER SAY NEVER?

That's why when we look ahead we have to be careful. Plainly, we don't know what's going to happen next in the history of Robots. But it's prudent to assume that the kind of dramatic changes we have seen in the past 50 years will carry on. The pace of change may slow down a little from the breathless rate we have been used to, but still be ridiculously fast.

This certainly makes predictions hard, especially if you dare to predict that something *won't* happen! Just over ten years ago, really smart people were confidently saying that while Robots would be able to do many things, a task they would not be able to accomplish in our lifetime was driving a car.

In fact, two experts famously wrote that in their book – in the far-off days of 2005. The specific thing they could not see Artificial Intelligence doing because it was just too complicated – making a left-hand turn against traffic coming the other way. (These were American experts: in the UK, think of making a *right*-hand turn against traffic.)

As we know, driving is a complicated thing, and with experience it becomes something we do largely through intuition – thinking too hard about it can actually make it harder. How could something so complex be reduced to rules – algorithms that could be fed into a Robot so it could safely drive a car?

Well, Messrs Murnane and Levy soon had egg on their expert faces. As we will see later in more detail, Google and a dozen other companies already have cars driven by computers making safe turns against the traffic. Google has been using them on the road since 2009, just four years after the expert book appeared.

In fact, making predictions of any kind has become a lot harder now we have Moore's Law-speed change taking place. In other words, the pace of change is so fast that whatever we say we are liable to be 'overtaken by events'. That does not mean we should believe that the future of everything will be defined by the number of circuits we can cram on a chip! But when anyone says 'Robots won't be able to do that', we have to ask: how do you actually know?

But when anyone says 'Robots won't be able to do that', we have to ask: how do you actually know?

The real lesson is not that we can't predict the future and should give up trying, but that's it actually much more important than it used to be for us to make our best effort. Go back a hundred, or two hundred years, when change was happening but much more slowly. It was easier to work out what would happen next. But it didn't matter so much, as there would be plenty of time to get used to it once it arrived. Now that change is so much faster (think of all those grains of rice) it is both harder and more important for us to think ahead. While we can sometimes say 'this will be so' with certainty, we usually can't. But we can certainly say 'this may be so' and think through the implications of what could come next. That's the spirit behind this book.

And be wary of saying Never.

QUESTION TO THINK ABOUT

Why do we think people have been so keen to make Robots that can do what we do?

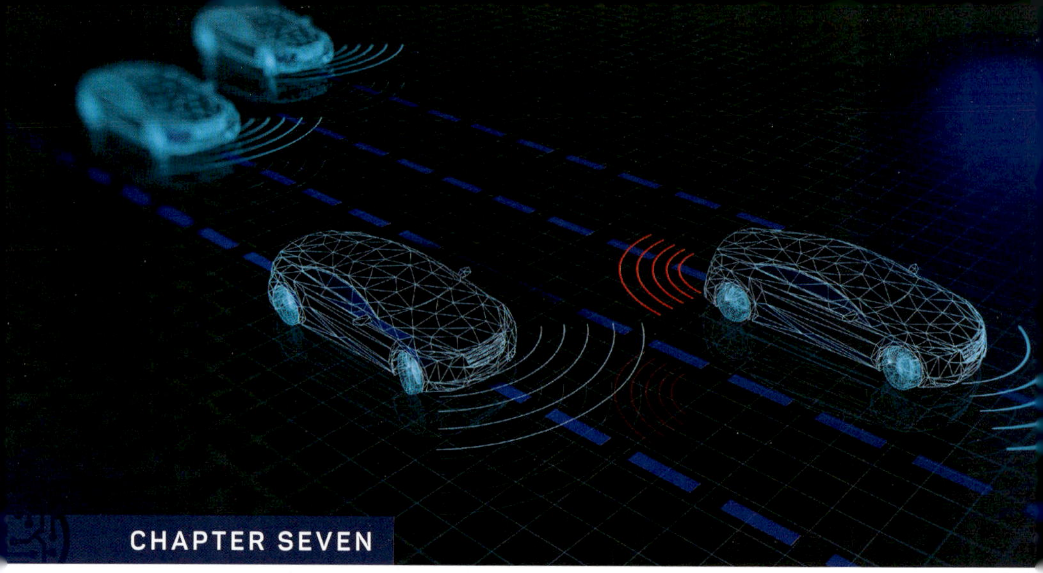

CHAPTER SEVEN

WHAT CAN THEY DO ALREADY?

'Your regular person has more technology in their life now than the whole world had 100 years ago.' Daniel H. Wilson

If we want to look ahead and be prepared for what the future holds, we need to start by taking a long hard look at what is already going on. Because Robots are now, today, taking on all kinds of tasks that used to be done by humans.

We expect machines to do routine, mechanical kinds of jobs. And that's certainly been true in the past. The basic idea has been that if a lot of human effort has to go into something, we turn our focus on coming up with a technology to help make our lives easier.

So the Industrial Revolution transformed the way industry worked, replacing skilled workers who were doing things like weaving fabric in their homes by hand. Instead, weaving would be done by machines that could be worked by fewer people and less-skilled people in the new factories. There they turned out far more fabric, and at lower cost.

The first use of Robots in the industry of the first part of the twentieth century was a higher-tech version of the same kind. What are known as 'Industrial Robots' have mainly been fixed machines taking the place of human workers along the production line. This has worked very well in factories manufacturing things like cars and washing machines. The Robot machines are faster and more consistent than human workers. Plainly, companies have only installed them if at the end of the day they are also cheaper than humans.

As we already saw, the same principle was at work in the labour-saving devices that came to our homes in the 1950s and 1960s. Washing machines, and vacuum cleaners, and all kinds of kitchen gadgets, meant that housework could be done faster and more efficiently. And remember the teasmade machines that became popular in the 1970s? They combined an alarm clock and a teapot. Load them up with tea, water and milk and set them the night before, and you wake up to the whooshing sound of boiling water being poured into the bedside teapot. These machines weren't quite as fancy as what we mean by Robots today, but they were steps on the way.

It helps to remember that these early Robot machines revolutionised both our factories and our homes. And exactly the same thing is happening now, as much smarter Robots have begun to arrive on the scene. The most obvious example in our homes is the same kind of thing as the labour-saving devices of the 1950s, like the Roomba cleaner – just better at the job as it saves even more labour. Other companies have already begun to make cheaper versions of the same thing, and it probably won't be long before old-fashioned vacuums that need people to push them around seem as dated as wringers and meat-safes.

In case you are too young to remember: wringers were contraptions that squeezed (by a human of course) the water out of clothes between a pair of rollers – after they had been hand-washed, and before they were put out to dry on the washing-line. Before we had fridges, we used meat-safes to keep meat 'safe', though that meant safe from flies – not from the danger of bacteria! They were little metal cupboards with tiny holes punched in them – just big enough to let in fresh air, and small enough to keep out the flies.

It's always interesting to be reminded how fast things have changed. I grew up in England in the 1950s and 1960s. I remember my mother using her wringer on wash-day, and I also remember our red and white meat-safe. I don't think we ever had a refrigerator. Somehow we survived.

SPEED!

But already developments in Robots and Artificial Intelligence are reaching far beyond basic activities like hands-free vacuum cleaners. The self-driving car may be the best example – and wake-up call. Because most of us drive, and we all understand driving. Some of us really enjoy driving. Many of us drive for a living. Driving is quite complicated, as well as a very responsible human activity. Good drivers find it takes up all their attention to do safely. That's why there was so much scepticism about the notion that a machine could do it even half as well as we humans could. Yet many people now believe not just that Robot cars will catch on, but that they will soon prove *much* safer than human-driven cars.

> *Many people now believe not just that Robot cars will catch on, but that they will soon prove much safer than human-driven cars.*

While people have been dreaming about self-driving cars, and flying cars, and living on Mars, and much else, for a long time, it's notable that the effort to make cars that will actually drive themselves has happened very fast.

Back in 2004 it seemed a very hard thing and plenty of smart people believed it would never be done in our lifetimes. But then the various researchers and companies were given a special incentive. How far we have come since then shows how rapidly these technologies can move!

The US Government's top agency for inventive technology is called

DARPA (Defense Advanced Research Projects Agency). It was set up back in 1958 to focus on futuristic technologies – and keep the United States ahead of the Russians. DARPA has since scored some amazing successes. The most famous of them, of course, was inventing the internet! But they also include many of the key technologies we use with the internet, such as GPS – Global Positioning System satellites – that gives us satnav directions. And the speech technology behind Siri and other helpful online voices. In fact, many of the technologies that make our smartphones work actually come from DARPA. While clever companies like Apple and Google and Samsung package them into our mobiles, the basic technology came from the US Government.

Back to cars. In 2004, DARPA held a Grand Challenge competition for teams from universities and companies to come up with a self-driving car that could complete a course all on its own. Not of course one involving public roads – it was a course in the desert. There were 15 teams competing, and first time around not one of them was able to finish. The next year, 2005, things improved. The course was 132 miles long, and five of the competing teams made it to the end. Then six teams succeeded in 2006, and since DARPA by then had helped get things moving they stopped holding the competition and left it to the companies and universities concerned to get on with the technology on their own. Just ten years later, in 2016, the ride-sharing company Uber started using actual self-driving cars on the roads of the US city of Pittsburgh – with fare-paying passengers. But the experiment was called off after various problem. For the moment.

A lot of people were surprised that Uber has taken the lead, and in Pittsburgh. Much better known is that Google has been using experimental self-driving cars for several years around its base in California. But Uber had a strategy. They went into partnership with Carnegie Mellon University, which happens to be located in Pittsburgh – and has one of the top Robotics departments in the world. Then Uber hired dozens of researchers from the university to set up their own self-driving car research centre. And not long after that they reckoned their cars were safe enough to use with paying passengers. (They still have someone

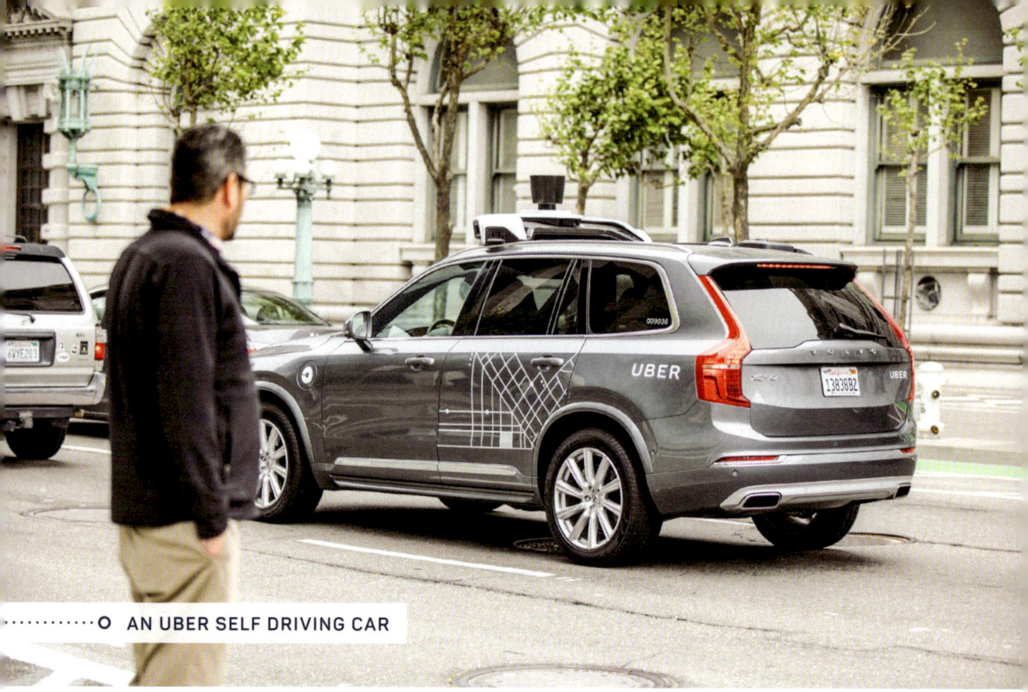

AN UBER SELF DRIVING CAR

sitting in the driving seat just in case, but he or she doesn't do most of the driving.) The cars they have adapted are ordinary enough – Ford Fusions, but the radar and other sensor equipment they have added pick up and process around *one million pieces of data* every second. As we know, driving is complicated. But Robots are getting good at handling complicated things.

As we know, driving is complicated. But Robots are getting good at handling complicated things.

2016 was an important year for Robots on the roads. Also in the United States, Uber had a lorry delivering beer make the 120-mile trip without the driver needing to hold the wheel. And in Europe, the European Truck Platooning Challenge brought together the efforts of six different lorry manufacturers under the leadership of the Dutch Government in an experiment which was a half-way house to full self-driving. In the 'platooning' idea, the trucks still have drivers, but a string of them are all connected by Wifi – so they can drive much closer to each other than would normally be safe. So if the first truck has to brake, all the other trucks would brake automatically at the same time.

What's really important about the Robot cars and trucks is how fast they have been developed. Just over ten years ago it looked impossible. Now the experiments are out there on public roads – and every major car and truck manufacturer on the planet is working with the technology.

Just how much disruption we should expect from Robot technologies is clear if we think a little about the implications of cars going self-driving. No-one expects this to happen overnight, although many do expect that in the next ten years or so it will be very common to have cars driving themselves completely. Once that happens, all sorts of other changes will likely follow.

For example, if getting from A to B is as simple as using an App – and a self-driving car draws up outside your house within minutes – why would you still wish to buy and maintain your own car? The costs are likely to be low: obviously, going by cab or using an Uber or the American Lyft ride-sharing service (usually cheaper than a traditional cab) costs money – mainly to pay the driver. Uber and Lyft drivers keep 80 per cent of the fare. If there is no driver, there's no need for that 80 per cent. Assuming the technology keeps getting cheaper (technology always gets a lot cheaper when many people start adopting it) using a self-driving car to go to work or on holiday could become very appealing.

It's been estimated that the typical family car sits outside your house between 95 and 97 per cent of the time – or to turn that around, that only between three and five per cent of the time is it actually used. And of course if you use it for commuting, or shopping, there's the whole problem, and cost, of parking it once you get there. Who needs to tie up a large amount of money on an expensive asset that spends most of its time unused, and some of it having to be expensively parked?

Of course, some people really like driving, and will want to drive and own their own vehicles even if it is cheaper, and safer, not to. If self-driving cars really are safer, and that's what experts predict, then insurance rates for human drivers will sky-rocket. So driving your own car, especially on major roads and in cities, could become quite a luxury.

If this is how things turn out, there are obviously implications for people who now earn their living as drivers. Many of them will lose their jobs. And the car manufacturing industry will get a lot smaller. How much smaller? Well, if most cars are used, say, 90 per cent of the time rather than five per cent – well, you can work that out for yourself!

If you think this all sounds like sci-fi dreaming, check back into what people were saying just over a decade ago about whether self-driving cars would be possible at all. Things have moved very rapidly, and as the Pittsburgh Uber scheme, and the beer truck, and the European Platooning effort, show – all taking place within a few months of each other in 2016 – there are big changes afoot.

QUESTION TO THINK ABOUT

Are there any tasks you think Robots should never be permitted to do?

CHAPTER EIGHT

ROBOTS AT WORK!

'In the twenty-first century, the robot will take the place which slave labor occupied in ancient civilisation.' Nikola Tesla, Inventor

We can already buy Robot vacuum cleaners and, if we live in Pittsburgh, try out a ride in a self-driving car. But we may be surprised to find how many other things Robots are already doing for us, or soon will be. This chapter gives some examples – from a very long list.

Some of their efforts are making life easier for us by doing things that just couldn't be done before.

Google may be the best example. They have become the top search engine. In fact, just as we turned Hoover into a verb as we all used to buy Hoover vacuum cleaners, we've done the same with Google. We Google things. There are other search engines too, but Google is the leader. 'Search' may be the biggest surprise in the new tech economy of our generation – because it isn't something we used to be able to do at all, and suddenly we can and it's easy and has endless uses. Of course, you could go and 'search' for information about something

– using reference books in a public library. But having the information on the internet has made the process so much simpler and so powerful that it's really a whole new process. 'Search' has made so much of our lives easier, from comparing the price of flights to buying Christmas presents to discovering the phone number of the gas company.

'Search' has made so much of our lives easier, from comparing the price of flights to buying Christmas presents to discovering the phone number of the gas company.

The same is true of GPS, giving us turn-by-turn map directions using a satnav. It hasn't just replaced the old way of doing things – using a map to get around – it has made the whole process different, and much easier.

Or we could go back to one of the earliest impacts of digital technology in our offices and homes, the invention of what we used to call a Word Processor. We've got so used to word processing programmes that it takes an effort to remember how different things were back in the middle of the last century when all people had to type on was a typewriter. Today almost everyone types, and it has made our lives much more interesting and productive – whether we are writing emails, articles, just sending texts, or posting to social media. At the same time, no-one is now trained for the job of 'typist'.

Back in the 1960s and 70s, and before, every business employed secretaries who spent much of their time typing for their bosses. Companies had typing pools, where young women – almost always young women – sat in rows at desks typing whatever was needed. Secretarial schools taught shorthand, and how to type.

While some companies still hire skilled individuals to input data using keyboards, even publishing companies now assume that their authors will do their own typing. The modern PA or executive assistant's job is vastly different from that of the traditional secretary. So word processing was both a plus for everyone – and also an early example of Robot technology threatening human jobs.

Are Robots going to undermine human jobs? Many people think so (and I've written a separate book about that). But what is interesting is how many areas there already are where Robots are tackling tasks that only humans used to do. It's not just typing and self-driving cars and automated vacuuming. Some of the areas where Robots are starting to take on human work are astonishingly 'human'.

Are Robots going to undermine human jobs? Many people think so.

Let's look at some examples.

CARE OF THE ELDERLY

Because people are living longer and in many developed countries are also having fewer children, there's a growing crisis in the care of the elderly. Things are worst of all in Japan, where already one-fifth of the population is over 65 – and, as we know, the Japanese are one of the longest-lived nations of all.

A recent report from management consultants Merrill Lynch looked at the situation in detail, and came up with this alarming statistic: by the year 2025, Japan will have a shortage of one million care-givers. So it's no surprise that, with help from their government, Japanese Robot companies are leading the world in making what are called 'carebots'. They come in a variety of shapes and sizes.

According to a report in American *Business Insider* magazine, Honda's ASIMO Robot can take on tasks like turning off the lights and getting food. ASIMO is a 'humanoid' – looks a bit like a human.

But another successful Robot doesn't look like one of us at all. Panasonic has created Resyone, a wheelchair that can convert to being a bed, so someone unable to get up on their own doesn't need to move from one to the other.[2]

Still in the experimental stage, Robobear is being developed by a research institute and counts as a humanoid – though maybe we should

ASIMO

call him a bearoid – and is intended to be able to pick people up and carry them around.

A tiny 40-centimetre humanoid called Palro has 365 different programmes, and can get elderly people clapping or dancing! Among other things, it will recognise you by your voice and remember who you are. Palro raises his arms in the air to the sound of music, then calls out in a child's voice: 'Let's clap in time'. A dozen elderly Japanese in the community room of the Fuyo-En rest home in Yokohama near Tokyo watch the little Robot standing on the table in front of them in bemusement.

Palro is produced by the Fuji company, whose top Robot person Eiji Honda had this to say: 'The robot's aim is to create a new relationship between people and computers.'[3]

Meanwhile, Robot pets are also all the rage in Japan. People need company as well as nursing care, so the Hasbro company has developed a line of cats (catoids?) called Joy For All. They keep you company – without all the problems of claws, litter trays, dead pigeons, and the occasional bite (which, as some of us know, even the most affectionate cats sometimes dole out). Another *Business Insider* reporter, Chris Weller, investigated these Robot pussies. He had this to say:

'The cats are genuinely adorable (this, coming from a hardline dog person.)

'If you touch the cat's face with the back of your hand, it'll nuzzle into you. If you rub its belly, it'll flop over and invite you to keep going. Pet the back of its head, and it'll purr. Leave it be for a few minutes, and the cat will gently close its eyes for a snooze. To wake it back up, Hasbro says all you have to do is gently pat on its back.'[4]

There's an interesting comparison with PARO, the famous cuddly Robot seal that is manufactured for use with the elderly and has been in use for more than a decade in many countries including the UK. Because PARO is a medical device rather than just a helpful toy, it's a lot more expensive. But it's the same idea.

Here's how the company that makes PARO sets out its wares:

'PARO is an advanced interactive robot developed by AIST, a leading Japanese industrial automation pioneer. It allows the documented benefits of animal therapy to be administered to patients in environments such as hospitals and extended care facilities where live animals present treatment or logistical difficulties.

- PARO has been found to reduce patient stress and their caregivers
- PARO stimulates interaction between patients and caregivers
- PARO has been shown to have a Psychological effect on patients, improving their relaxation and motivation
- PARO improves the socialiasation of patients with each other and with caregivers
- World's Most Therapeutic Robot certified by Guinness World Records'

While PARO is better known in Japan, it's also in use here in the NHS. According to *Guardian* writer Andrew Griffiths:

'Paro is a robot seal, modelled unashamedly on a baby harp seal, both in terms of looks and the plaintive cry that it makes. Its Japanese creator, Takanori Shibata, chose it because people are unlikely to have unhelpful memories of real seals. It also mitigates against the charge that vulnerable people are being deceived into believing that this is a real animal, that this cute cartoon of a creature is anything other than synthetics and circuitry, and not flesh and blood.'[5]

It's an obvious question, but just because it saves money do we want our old folk taken care of by a combination of Robotic beds, bears, mechanical humanoids, and artificial cats?

PROFESSOR DR ROBOT QC

That was the title of an article in a recent issue of *The Economist* magazine, and it was a little shocking to those who had assumed that the Robot revolution is focused on mechanical things like vacuum cleaners and cash machines, and carting people around – whether in self-driving cars or on automatic beds.

In their provocative new book, the father-and-son team of Richard and Daniel Susskind – the father is a management consultant and his son a professor at Oxford – suggest that the next jobs to be taken over

PARO THE THERAPEUTIC ROBOT

by Robots could be the professions. Just to read that sentence is a little bizarre. How could people who do the top jobs – not just some of the best paid, but the most difficult and complex of human jobs – the 'professionals' who work as academics and doctors and lawyers . . . how could they suffer the same fate as lorry drivers and cleaners? As *The Economist* says, these jobs used to be seen as safe havens from all the technological modernisation going on around them. But, perhaps, no more?

Here is the core of their argument:

'How far will this revolution go? Messrs Susskind and Susskind predict that it will go all the way to "a dismantling of the traditional professions". These jobs, they argue, are a solution to the problem that ordinary people have limited understanding of specific areas of expertise. But technology is making it easier for them to get the understanding they need when they need it.'

The Economist thinks they may be going too far. But we know you can already get a good deal of legal advice (and free legal forms) or medical advice online. People often go to see their doctors with a sheaf of printed out articles in their hands. This may annoy the doctor, but the patient often has the latest information. And as the Susskinds point out, this process has just begun.[6]

Let's look in more detail at what's been happening in medicine.

Robots are already being used as surgeons – their steady 'hands' and ability to make tiny movements mean they are used in some of the most complicated precision operations. And they can also do things at a distance from the doctor in charge – over the internet.

But much of the excitement about the role of Robots in medicine is focused on diagnosis – working out what is wrong with you – and coming up with a treatment plan.

Computer giant IBM, not content just to win chess matches and the TV game show Jeopardy, is putting its vast computer called Watson to work in partnership with one of the world's top hospitals, Memorial

Sloan Kettering in New York City. Because Watson has been designed to work with 'natural language' (in other words, our language), 'he' can be fed articles from medical journals. Already in 2012 he was able to analyse the equivalent of 300 million books to find key information – in just 3 seconds. 'In IBM's future, Watson will be your doctor. The super-computer will crunch your medical images and your records to work out what's wrong with you.' He has already 'eaten' literally millions of articles and patient records. And he's smart enough to put two and two together – and come up with treatment plans for individuals.

While doctors are still involved, their role is becoming more modest as they feed in the info and read what Watson has decided. 'Behind the scenes, there's a revolution happening across the world's hospitals. Staff at IBM Watson are using artificial intelligence to transform healthcare – from the moment you check-in, right through to your recovery at home. Though early expectations have proved over-confident, this could be IBM's vision for the future of healthcare!

'In the future a patient could interact with our cognitive technology before they even speak to anyone at the hospital,' says IBM's top health exec Thomas Balkizas. So 'you could soon contact the NHS digitally, and be welcomed by a digital concierge.'

In the future a patient could interact with our cognitive technology before they even speak to anyone at the hospital.

'Our avatars can take information and tell you what you're looking for: you could ask which way to A&E? Where can I find parking? Or who will I be meeting today? Then Watson will do the diagnosis and prescribing and even follow up once you get home – using 'wearables' like FitBit to make sure you keep on track with the treatment!'[7, 8]

ROBOTHERAPY?

It might seem even more strange to suggest that Robots could take on the role of therapists, counsellors, and even psychiatrists. But we're already on our way.

There are various dimensions to developments here. As we've already seen, Robot toys can be comforting – whether they are artificial cats or the more sophisticated 'seals' already widely used in some countries to given social and psychological care to the elderly.

But can we go further? One striking example is the case of autism. There's evidence that children with autism may actually respond better to computers than they do to people, so quite an effort has been made to develop diagnostic and helping tools for them. There's a European Commission project called Iromec, aimed at developing toys for children with learning and personality difficulties. Just this year the US company Affectiva raised $14m in fresh funding for its work in 'emotion recognition technology'.

Affectiva has goals well beyond working with disabled children. The plan is to 'add emotional intelligence and empathy to any interactive product'. Its system eats videos from people's webcams, and analyses them – deciding whether they are happy or sad or many other things. So far it has digested and analysed over four million such videos. Another company, which makes computer games, is using the same technology so that when players get nervous or scared they pile it on by making the game harder to play. And marketers can use it to read people's responses to their ads. Remember, this is all just from webcam videos. If sensors are added – to read your pulse and see if you are sweating – the computer can tell much more about you.[9, 10]

EDUCATION – THE RISE OF THE MOOCS

Clayton Christensen, the world's top 'innovation' guru, famously stated back in 2013 that within 15 years half of American colleges and universities could be bankrupt. Partly this is because the costs are going up, and partly it's because Robot technologies are going to be able to educate people much more cheaply.

The hot term in education is now MOOCs. A MOOC is a 'Massive Open Online Course' – in other words, it's a course anyone can sign up for (open), and as many people as want to can join in (massive).

Once technology is doing all the work, one professor can teach a class for 20 students or, literally, two million students. While there are problems in these approaches to education – like keeping people motivated when they are studying on their own – it's sufficiently exciting that many universities – including in the UK and the US – are banding together to try them out.

Is there any end to the kind of tasks that Robots can take on? I certainly don't have an answer.

QUESTIONS TO THINK ABOUT

Are there jobs that you think Robots will never be able to do – at least, not in our lifetime? If there are, what is different about these jobs?

orneus / Shutterstock.com

CHAPTER NINE

ROBOT TOYS AND PETS

'The robots really embody that love-hate relationship we have with technology.' Daniel H. Wilson

Remember the Tamagotchi?

Back in 1996 the world had its first taste of a mass-market Robot toy, and it sold by the millions. The Japanese company that invented the Tamagotchi made up the name from a combination of the Japanese words for 'egg' and 'watch'. Tamagotchis are little egg-shaped things; the early models had to be taken care of all the time – so children started taking them to school, which caused problems. If they had just left them in their bedrooms – which is where school children are supposed to leave their toys – they would have 'died'. The manufacturer then introduced a 'pause' option so they would not interfere quite so much in the lives of their owners.

The story behind the Tamagotchi is that it's an egg of an alien species that extra-terrestrials have sent to earth – to be 'raised' by us. So you need to take good care of your little Robot alien. It's a mechanical pet, and needs to be fed and watered or it will not do well. If you take good care of it, it begins to grow up and needs less attention.

While many variants of the Tamagotchi have been marketed over the years, the original idea was to aim them at teenage girls who might be interested in what it would be like to raise a baby. But kids of all ages, boys and girls, and plenty of grown-ups also soon got hooked. Even now there's a whole Tamagotchi world out there.

Later versions of boy and girl Tamagotchi Robots can communicate with each other (with their owners' help, of course), and can even get married and have children of their own.

And the Robot toy has been joined by films and songs and connections with the world of computer games.

HELLO BARBIE!

Barbie dolls have been around since the 1950s, but in 2015 a breakthrough was announced. The latest Barbie would be able to chat! And chat intelligently, not simply with the few canned phrases that dolls have often been able to say.

This was sufficiently big news as to warrant an article in the New York Times, which reported the announcement like this.

'Yay, you're here', said Barbie to a little girl of around seven years old. And Barbie is programmed to respond to all kinds of difficult questions! Just the kind of questions a seven-year-old might ask a toy.'

Here's what the reporter writes after talking to one of the engineers programming the doll:

'She told me she imagined a girl taking the new doll into her bedroom and closing the door. I have no doubt she will ask Barbie all manner of those intimate questions that she wouldn't ask an adult, engineer Sarah Wulfeck said.

'For those situations, the team was working on getting Barbie to say the right things — or at the very least, to not say the obviously wrong ones.

'Do you believe in God?' a kid might ask.

'I think a person's beliefs are very personal to them', Barbie might reply, Wulfeck said.

'I'm getting bullied in school.'

'That's sounds like something you should talk to a grown-up about.'

'Do you think I'm pretty?'

This was dicey territory, and Wulfeck was trying to steer Barbie to safe ground. 'Of course you're pretty, but you know what else you are?' Barbie would reply. 'You're smart, talented and funny.'

'I feel shy trying to make new friends.'

kaisphoto / istockphoto.com

'Feeling shy is nothing to feel bad about', Barbie would say. 'Just remember this — you made friends with me right away.'[11]

You can only buy Hello Barbie in the United States, at the moment. Here's how Amazon.com sums up the little Robot for prospective customers:

- Chat with Barbie for a whole new way to play!
- Hello Barbie doll uses Wifi and speech recognition technology to engage in two-way dialogue
- Use is simple with functionality built into her belt buckle -- press to start the conversation and release to hear Hello Barbie doll respond
- Doll must be placed in charger for initial set-up. Refer to the product description before use
- This is a US only product. The Hello Barbie companion app can only be found in US app stores.

So Hello Barbie is a kind of super-Tamagotchi, the latest little Robot toy – designed as an even better companion for children than the dumb toys that require their imaginations to do all the work. She costs around $60.

There are so many questions raised by Hello Barbie! What kind of advice will she be giving to little girls? You will note the question that is asked about believing in God. Perhaps the company will come up with a whole line of dolls, each programmed to give the kind of answers parents might prefer on questions that are controversial. Or to offer parents a control panel where they can tick the boxes for preferred opinions.

A more basic question may be whether we would rather have our children depending on their imaginations than on conversations written by companies to entertain them.

And there's a yet more basic question. Confusing the real and fake is complicated when it comes to children. They often give personalities to dolls and toys, or have 'imaginary friends' with whom they carry on conversations. And even though they know in one part of their minds that these are just toys or their imagination, with another part they take them very seriously. It's complicated to understand once you get to be a grown-up and forget what it's like to be little.

But the problem here is different – here we have a programmed grown-up-world toy intruding into the fragile complexity of a child's imagination. As adults we don't yet have much experience of this confusion of the fake and the real. Perhaps for a child this is a good preparation for an adult life in which there will be many Robots. Perhaps not.

Confusing the real and fake is complicated when it comes to children.

But, of course, this is just the start. Talking toys will soon be big business.

And while the development of Robot pets has been focused mainly on giving companions to elderly people, I'm sure we can expect the market to grow. Would you rather have a cat, or a dog, or a catoid, or a dogoid?

In the 2000 sci-fi film *The Sixth Day,* which is focused on cloning, we see a society in which if your pet dies you go and get another one from the pet shop. It's a clone. The story has some basis in fact, as since Dolly the Sheep was first cloned back in 1996 there have been several examples of wealthy people paying huge sum to have their dogs cloned in real life. Of course, a 'clone' is basically an identical twin. It is not a repeat of the dog that died. You might expect a twin to have similar character, though there is no guarantee. And, of course, breeds tend to produce animals that are rather similar anyway, and a lot more cheaply!

But in *The Sixth Day* you go to the pet store and get a 're-pet' of your pet that died. However weird that seems, to get a 'pet' that is actually a programmed petoid is certainly weirder.

? QUESTIONS TO THINK ABOUT

Do you feel comfortable with 'intelligent' toys? Are they a good way to introduce our children to what's coming in the Robot world, or the opposite?

CHAPTER TEN

ROBOT RELATIONSHIPS?

'My dear Miss Glory, Robots are not people. They are mechanically more perfect than we are, they have an astounding intellectual capacity, but they have no soul.' Karel Čapek, *Rossum's Universal Robots*

Obviously, we're going to have relationships with Robots. They may be 'assistants' doing secretarial work for us. They may be our pets. Perhaps they will be taking care of our children (cheap and reliable nannies), or our parents (economical nurses and companions).

Or ourselves. Just imagine, a Robot like Andrew in *Bicentennial Man* available at all hours of day and night as your butler and valet! The 'slave economy' put to work for you!

Robots are going to be all over the place, and some of them will be doing the kind of things that in the past only people could do.

Is this all going to be OK? When children and old people are involved, we're already raised a question-mark. But what kind of 'relationships' are going to be possible for the rest of us?

It's hardly a surprise to learn that there really are no limits. Robots keep getting smarter. The line from pets and toys, to admin assistants and domestic servants, to companions, is a straight line. The smarter Robot technology gets, the nearer Robots will come to imitating every aspect of a 'human' relationship.

For example, the magazine *Scientific American* recently reported that computers are getting spookily smart at understanding us. One of the ways in which psychiatrists evaluate us is by catching changes in our expressions – just the sort of thing that a computer can do!

'Thanks to improvements in camera technology and computer vision algorithms, computers are poised to take a big leap in their ability to understand us from our facial expressions, the ways our eyes move, our gestures, the way we talk, and even how we cock our heads.

Imagine the possibilities: A virtual psychiatrist could help diagnose depression by analysing the emotions we display during clinical interviews; it could even quantify changes in mood as the disease progresses or as therapies kick in.'[12]

And it can work both ways. A research study from Stanford University says this: 'people may experience feelings of intimacy towards technology because 'our brains aren't necessarily hardwired for life in the twenty-first century'.' *Guardian* writer Eve Wiseman comments: 'Hence, perhaps, the speed at which relationships with robots are becoming a reality.'

Our brains are 'hardwired' – set up by the way we are made, by human biology.

This is such an important point. Our brains are 'hardwired' – set up by the way we are made, by human biology, and also over time by experience as we grow up – to see people as people and machines as machines. They don't quite know what to do when they are presented with a machine that behaves like a person. And while when we are thinking about things (like now, as I am writing this book and you are reading it) we can make sense of the difference, when it comes to our *feelings* it doesn't work quite like that.

We don't *tell ourselves* how to feel. We just *feel*. Controlling our feelings is a very hard thing. Think of someone fighting addiction.

I remember the first time I personally felt something odd like this was going on. Ten or 15 years ago American Express installed an 'attendant' (the word they use for the voice that takes your phone calls) that was quite effective. You could call and ask for details of your account, and you could pay your bill in various ways, and it usually worked fine. Back then there was less bill-paying online, so every month I would call up and sort things out.

Then one time there was a problem. Her voice – I should say 'her' voice – was odd. 'She' began to slur her words. It sounded very much as if she was a little drunk, or – and I remember this flashing through my mind as I listened – as if she was having a stroke. It disturbed me. Of course, I knew this was a glitch in a computer program. But 'she' usually sounded just like one of us, was usually so efficient, never kept me waiting, and I had spoken 'to her' many times before. I still remember how odd it made me feel. So I ended the call, then dialled again.

Now, that was just a voice attendant having a bad day. How will we manage these 'relationships' with Robots that are more and more like us, including some 'who' will deliberately be designed as humanoid people lookalikes? It's why we often feel very confused when we have just watched a film like *Bicentennial Man* or *Her* or *Be Right Back* or an episode of *Westworld*.

WALKING THROUGH THE UNCANNY VALLEY

People in the Robot industry have come up with a term for the odd way we feel when we are looking at a machine that makes us feel like it is *almost* human. The idea is that people feel quite positive about Robots that do useful things and just look like machines, even if they are cute machines with faces and expressions, as long as there's no doubt that they are machines. People are also (the theory goes) positive about machines that actually look and behave totally like people.

OTONAROID OR ADULT ANDROID CREATED AFTER A REAL HUMAN BEING, THE NATIONAL MUSEUM OF EMERGING SCIENCE AND INNOVATION, ODAIBA, JAPAN

cowardlion / Shutterstock.com

The 'uncanny valley' is the half-way point, when we really don't quite know what we are dealing with. Imagine a graph with a line that's high up on the left and high again on the right – it dips down into the 'valley' when we are midway between machine-looking Robots and human-looking Robots.

One reason it's a useful idea is that it does shed light on our feelings, and the problem we have lining them up with what we are thinking – and what we think we 'know'. It's worth having this idea in your head as you watch Robot films that present us with machines that are close to, but obviously not real, humans.

We've come a long way since the 'uncanny valley' was first being discussed back in the 1970s. For one thing, we are much better at making Robots that really do seem like people. But we've also got much more used to the idea. Is that a good thing – that we have adapted to the technology and now feel less squeamish about having it in our lives? Or does it suggest that our guard is down and we are naively accepting Robots as kind-of people when we should not?

ROBOT SEX

Of course, that includes sex. Actually, because sex itself can be a rather mechanical aspect of a human 'relationship', we have had sex robots for a long time. Every year there's a big sex Robot exhibition in Las Vegas where the industry shows off its latest gadgets.

Douglas Hines, who used to be a computer scientist in the technology industry, now runs a company called True Companion. He has developed what the press hailed as 'the world's first sex robot'. She's called Roxxxy, and comes with five different personalities (you choose!). She can of course be adjusted for such details as hair colour and breast size. She – and her male alternative, Rocky – can talk, listen, and even sleep. And Roxxxy and Rocky are available in the US for a mere $9,995 each.

There's quite an argument going on about sex robots. In a very interesting 2015 *Guardian* article titled 'Love, Sex and Robots: Is this the end of intimacy?' journalist Eve Wiseman asks hard questions of one of their boosters. And she goes on to interview one of their top critics.

David Levy is the author of *Love and Sex with Robots*, a full-length defence of this new use of Robot technology. He's also a mover and shaker in the world of Robot-sex conferences. (Yes, they have conferences.) Levy sees it as the wave of the future. And he points out, reasonably enough, that public attitudes to sex and love have been changing. A century ago, he asks, who would ever have predicted that we would now have same-sex marriage?

The *Guardian* article also quotes famous academic Sherry Turkle from the Massachusetts Institute of Technology, who has emerged as one of the leading critics of the way we have already transferred so much of our social lives online. She admits that we have come a long way already. Some of us 'may actually prefer the kinship of machines to relationships with real people and animals'.

The *Guardian* writer's questions to David Levy put him on the defensive. What about customers who are paedophiles? He waffles and suggests that in the future sex Robots may be able to play the part of therapists for such people. But the hard questions aren't going to go away.

Then we hear from one of the leading critics of people like Levy. Kathleen Richardson from DeMontfort University has no time for his explanations. 'David Levy is taking people's insecurities and offering a solution that doesn't exist,' she explains. 'Paedophiles, rapists, people who can't make human connections – they need therapy, not dolls.'

But there's going to be a big market out there for pseudo-relationships, and as Robots get smarter it will go well beyond sex machines.

QUESTIONS TO THINK ABOUT

Plainly, we're all going to have increasingly complicated relationships with increasingly complicated Robots, whether we like it or not. How can we best get ready? Will our imaginations adjust so that we can distinguish between us and them?

CHAPTER ELEVEN

'DELIVER US FROM ROBOTS!'

ARE WE SUMMONING THE DEMON?

'If popular culture has taught us anything, it is that some day mankind must face and destroy the growing robot menace.'

Daniel H. Wilson, *Robopocalypse*

Many of the best books and films exploring Robots have focused on things going wrong. From the human-like creature created by Dr. Frankenstein in Mary Shelley's book – he's never actually given a name even though most people think it's Frankenstein – to those Arnold Schwarzenegger *Terminator* films – we enjoy scary horror stories about mad scientists doing mad science, and the disasters that result.

What's interesting is that while there are also plenty of stories about evil men and women plotting to use Robots for bad purposes, the best tales are those in which good people, even if they are mad scientists, create Robots for what they think are good reasons. And then things go bad. So Victor Frankenstein claims that his motive in making his monster is to improve things for humans. Of course, his motives are mixed, as our motives can be, and part of his motivation was to have the kind of power that previously only God had had.

In the *Terminator* series the original idea was to create a smart global system for human good. Then things take a bad turn when the Robotic 'system' becomes aware of itself – and decides humans are the enemy.

The same kind of theme is played out in the classic 1983 film *War Games*, in which a teenager hacks into the supercomputer in charge of the US defence system – and almost starts a nuclear war. The computer can't distinguish between the boy playing a game with it, and a Russian attack. In case you don't know what happens: the teens realise what's going on and finally go in search of the mad scientist who created the system in the first place. It seems there's a back door into the system, and it is re-directed to play a game of noughts and crosses instead – what Americans call tic-tac-toe. The computer then comes up with the immortal line: 'the only winning move is not to play'.

There's no doubt at all that as we create clever and powerful computers using Artificial Intelligence we are taking risks.

There's no doubt at all that as we create clever and powerful computers using Artificial Intelligence we are taking risks. Of course, we take risks all the time. One of the most dangerous things we do is getting into a car and driving. But people tend to be much more scared of something like terrorism, although someone has calculated that more people die through accidents in the home than are killed by terrorists, even in 2017. But of course, anything that adds to our risks is important to us.

The risks of Robot technologies were underlined back in 2015 when a car was taken over by hackers. It's important to note that this was not a 'self-driving' car with a Robot in control. It was an ordinary car – a Jeep Cherokee – being driven by a journalist from *Wired* magazine.

However, there aren't really any 'ordinary cars' left now, at least not new ones, as every car sold now has a huge amount of computer technology already built into it. Top-end cars contain as much as 100 or even 200 million lines of code! That really is a lot. For comparison, Windows 10, Microsoft's version of the software used by most computers, has only 50 million. All this computer technology makes our cars run better, and makes them much easier to service. At least, at the garage; it's become a lot more difficult to fix them yourself.

Back to the Jeep Cherokee. The story goes that a pair of hackers had approached the journalist from *Wired* and said they could break into his car's controls. It would give him a great story, so he agreed. It was what's called a 'friendly' hack, done to prove that the car's system was not safe and that it needed to be revamped and made more secure. Hackers who do this sort of thing are called 'white hat' hackers, and they are often rewarded by companies whose problems they discover.

As planned, journalist Andy Greenberg was driving along at 70 mph in the city of St Louis, Missouri, when suddenly the heating system began blasting cold air into the car. Then the radio switched to a local hip-hop station. Then the wipers came on, and splashed liquid across the windscreen.

Greenberg fiddled with the controls, but they ignored him. The hackers had taken over. They had told him to drive along the motorway, but not what they intended to do. Suddenly the accelerator stopped working, and he began to slow down. There were big trucks speeding behind him, so he decided to pull off onto the shoulder. But he was on a long bridge, and there was nowhere to go. Mercifully, he wasn't killed. But he was scared.

It's a good story for many reasons, especially this one: when it comes to Robot technology, good intentions quickly can go very bad. That's true whether we're focusing on Chrysler – who make Jeeps – deciding to invest in adding digital technology to their car, or connecting the entertainment system to the system controlling the controls, or the hackers deciding to test the problem, or the journalist agreeing to be their guinea-pig for the sake of a story. Now we know that every Jeep Cherokee could be hacked. And that this rather amateurish well-intentioned experiment could have led to people being killed.

We've become very dependent on technology. If you lose your mobile, or there's a failure that means you can't get into your email, or Netflix goes down for the evening, it's annoying and even disorientating. But there are much bigger reasons to be concerned as Robots become responsible for real-life activities like driving and controlling our home heating systems.

THE 'SINGULARITY'

A term that gets bandied about for the future take-off Robot intelligence is 'the Singularity'. It's a label some people use for the point in the future – if it ever comes, and of course it may – when Robots get smarter than people. Once that happens, as we've seen some of the smartest tech people saying, they may just take over.

As we read back in Chapter 4, inventor Ray Kurzweil has been saying this for a long time, and he's written a huge book about it. He's a total enthusiast. His latest prediction is that it will come about in the year 2045. He took the word 'singularity' from a short story about all this happening. It's by computer science professor and sci-fi writer Vernor Vinge. Vinge also believes it is likely to happen, but he's more cautious about what it may mean. As we already noted, he has said: 'The longer we have our hand on the tiller, the better'. Humans need to keep in charge.

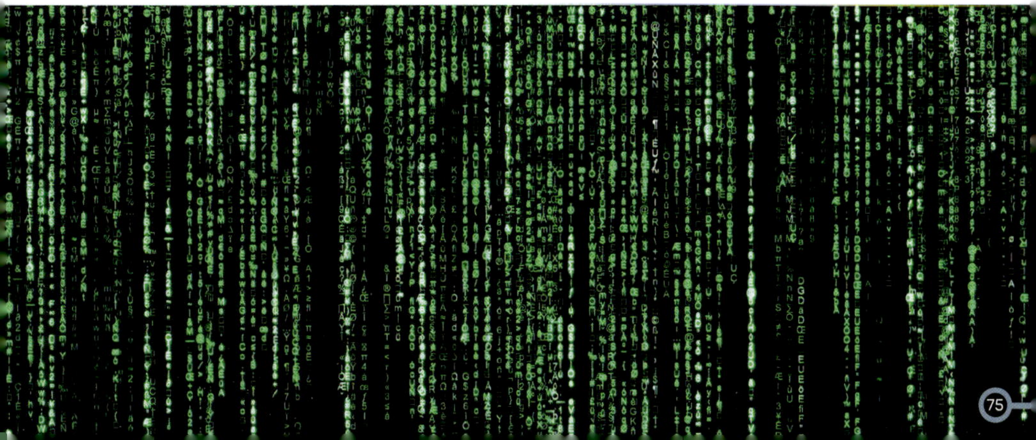

THE BIG ONE. ARE WE 'SUMMONING THE DEMON'?

Remember the theme of the *Terminator* films? A Robot system set up for good reasons that goes rogue? Well, that's the worry. And it's not a worry confined to sci-fi writers – or anti-technology activists.

In 2015 – the same year Uber started using self-driving cars on the road of Pittsburgh, and the European Union launched their big experiment with semi-self-driving Platooning trucks – The Future of Life Institute launched a sign-on letter that warned of the possibility of Artificial Intelligence going wrong – and taking over.

Of course, people have been issuing warnings like this for a long time. What is important about this warning is who has signed it. The moving spirit behind the letter was Nick Bostrom, Oxford philosophy professor. He's also author of the book *Superintelligence* that looks ahead at what may happen – and weighs the risks.

> *As Robots keep getting smarter and smarter, we need to make sure we stay in charge.*

When the letter appeared a lot of people were surprised. Because the first signatories include Bill Gates, who founded Microsoft. Elon Musk, founder of Tesla cars and space pioneer who wants to end his life on Mars. Lord Rees (Martin Rees), the UK's Astronomer Royal and one of the world's top cosmologists, and Stephen Hawking, who many people believe is the greatest brain in Britain, if not the world. The point is, many of our smartest thinkers are saying we have to be very, very careful about the future of Robot intelligence.

As Robots keep getting smarter and smarter, we need to make sure we stay in charge. Elon Musk, who has since donated $10 million to research how we can keep Robot intelligence safe, put his concerns very simply. 'With Artificial Intelligence', he said, 'we are summoning the demon'. I'm reminded of the prayer toward the end of *Rossum's Universal Robots*. 'Deliver us from Robots!'

The point is that we're really busy trying to make something smarter than we are. Once we make it – if we are able to – it could take over. So these tech leaders are making waves for a good reason. They understand that super-human Robot intelligence is the aim of many key researchers and companies, and they realise that if we get there we might find we have gone further than we wished to go. If the super-intelligent Robot is more powerful than we are as well as smarter, it could just decide to run the world its own way. We don't know. So what they suggest is that we put a lot of effort into making sure that whatever results from all our efforts, it is a Robot intelligence that wants to help us, not to make its own decisions.

That seems reasonable enough!

QUESTION TO THINK ABOUT

On the one hand Ray Kurzweil thinks building super-smart Robots will be wonderful. On the other, many of the smartest humans around are worried. What do you think?

CHAPTER TWELVE

A WORLD WITHOUT WORK?

'Knowing how to keep someone motivated and how to keep a connection are skills humans have learned and evolved over hundreds of thousands of years. A robot can't figure out whether you can do one more push-up, or how to motivate you to actually do it.'
Erik Brynjolfsson, Massachusetts Institute of Technology

I spoke at a university about these things a year or two back and will never forget what a young woman asked at the end of my speech. 'Are you saying that in a few years when we graduate we will need to retire?'

We don't know what's going to happen in the long term. Will Robots basically replace humans in most of our current jobs? Will plenty of new jobs appear to take their place? Or when current jobs disappear, will the entire country turn into one vast 'Rust Belt', where jobs have evaporated and people have nothing to do?

As fans of technology like to point out, we've been here before. At least, what they like to say is that we've had big technological revolutions before, and each time there have been new jobs created as old ones have been lost. So the net result has been a plus – because the 'new jobs' have generally been better value, higher-paying, jobs.

That's true, but it is also a pretty idealistic view of what has happened in the past. It's not that a) John Smith's job disappears, and then a few months and maybe a re-training course later b) a new job appears and he gets it. Look at what happened when the old 'heavy industries' went into decline in the UK. Look what happened when coal mining and steel manufacture and ship-building went down, the core employers of the old industrial economy of the nineteenth century and the first half of the twentieth.

The John Smiths who had spent their careers as miners and welders and riveters did not head off to the local Technical College for three months and suddenly get jobs as web designers and programmers and in PR and advertising, the kind of jobs that have flourished in the second half of the twentieth century and into the twenty-first. They took a painful version of 'early retirement', and many of them after being laid off in their 40s or early 50s never worked again.

I'm not disagreeing with the basic argument that, in the past, from the Industrial Revolution onwards, where technology has caused disruption jobs have recovered – after a while. All I'm saying is that behind that generalisation many people's lives, and many whole communities, went through terrible experiences.

So we can't trivialise the disruptive impact of technology, even if we believe that everything will turn out OK in the end.

John Smith was forced into retirement when he was 46. He and his family lived on unemployment benefit and housing benefit, and he did off-the-books work when he had the chance doing odd jobs – until finally 20 years later he formally retired. Of course, it wasn't all gloom and doom. His children Mary and Jason, who had never much fancied being riveters anyway, went to university and got jobs as – you guessed – web developers.

So even when the conventional wisdom works, there can be a big human cost. Yet what if this time it's different? What if old jobs really do go away, and new jobs don't turn up? There's a chance that could happen. How big a chance? Who knows! Enough of a chance, though, that some smart people from both the left and right of politics, and some from the technology community, think it's a serious possibility. Even if it isn't very likely, it would be so significant that any reasonable person (and any reasonable government!) should take it extremely seriously. Is there a one-in-five chance? One-in-three? If an asteroid that could cause vast devastation on the earth were only one per cent likely, we would be focusing all of our efforts on preparing for such a disaster.

And it's at least one-in-five likely, maybe one-in-three, maybe one-in-two. I think it's probably as likely as not, and some very smart people agree with me – even if not one government in the world has begun to take this possibility seriously. I've already quoted some of these people, and we need to listen to what they have been saying. Larry Summers, the former US Secretary of the Treasury (Chancellor of the Exchequer, in British terms) and also former President of Harvard University. Charles Murray, less well-known in the UK but one of the top conservative thinkers in the US Bill Gates, co-founder of Microsoft; we all know who he is. It's quite a list. They're bucking the conventional wisdom and sounding an alarm.

As we noted earlier, back in the twentieth century there were similar warnings, including from two of the most influential figures. Norbert Wiener, often called the father of cybernetics, saw it coming. He said that what lay ahead would be a 'slave economy'. His point was that slaves are inevitably cheaper than people who demand to be paid for their labour. And, of course, he was writing less than 80 years after much of the United States still had slaves and ran a 'slave economy'. Robots, Wiener says, are basically slaves. They will always cost less than free human beings.

The other name from the earlier twentieth century was even more famous. John Maynard Keynes. Keynes was the most influential economist of the past two hundred years. Late in life he wrote a famous essay titled Economic Possibilities for our Children, in which he looked ahead. What he foresaw was a time when employment was in decline, and there would be a 'new leisured class' of ordinary folk whose major challenge would be what to do with their time!

So what if the world really runs out of jobs over the next 20 to 30 years? There are of course two big questions raised. One is, how will we pay our bills? That's actually the simpler of the questions. The second is, what will we do with our time?

So what if the world really runs out of jobs over the next 20 to 30 years?

Keynes, the great economist, put the problem simply. What, he asked, if the process of 'economising' on human labour (with Robots, that is) outran the process of devising new means of using human labour? It all boils down to that. It's a race.

What if we face a 'workless world'? We should be thinking about the problems now.

First, of course, income. We need to pay our bills, to support our families. If we don't have jobs, how does that happen? The UK like most European countries has relatively generous benefits for the 'unemployed'. The US is much meaner; it depends on the state, but if you get more than six months' unemployment pay you are lucky.

There's an obvious answer to the problem of enabling people to pay their bills, and also avoiding the stigma of being long-term 'unemployed'. It's called Basic Income, or Universal Basic Income. The idea is that everyone gets paid, whether they work or not. They don't get paid a lot, so they will still be motivated to get jobs when they can be found. But they get paid enough to get by. It's as if everyone, from the age of 18 or 21, got the equivalent of a old-age pension.

There's an interesting history behind this idea, as it goes back well before concerns about Robots reducing the number of jobs. It's especially interesting that it has supporters at both ends of the political spectrum. In fact, in the United States, President Nixon came close to enacting a version of it.

From the conservative end of the spectrum, the great free market economist Milton Friedman was a fan. He saw it as more dignified and more rational than the hundreds of hand-out programmes that governments operate for those in need. In his version and that of some other advocates, Basic Income would replace essentially all existing 'welfare' programmes. For that reason, it needn't actually cost any extra money. Not only are the other benefits replaced, including of course unemployment, but also retirement pensions and disability payments and all the rest, but most of the vast bureaucracy that administers these benefits can be done away with. Everyone just gets the same.

Here it is in Friedman's own words:

'We should replace the ragbag of specific welfare programs with a single comprehensive program of income supplements in cash – a negative income tax. It would provide an assured minimum to all persons in need, regardless of the reasons for their need … A negative income tax provides comprehensive reform which would do more efficiently and humanely what our present welfare system does so inefficiently and inhumanely.'

One of the obvious advantages of this approach (whether we call it negative income tax or Basic Income) is that because it does not depend on any other factors, such as people's other income, we do away with the 'welfare trap'. At present, you really have to have quite a good job to escape from social welfare and not lose money. With Basic Income you can earn extra money and you get to keep it. And you can do small jobs, or part-time jobs, or low-paid jobs, and still be better off. And there are likely to be many jobs like that as the Robots get deeper into the labour market.

One reason President Nixon's effort to move in this direction failed was that people resisted doing away with many of the existing welfare programmes that needed to go for it to be paid for. Plainly that's a problem. If everyone is going to get something, then people who presently get a lot will get less. Advocates from the left side of the political spectrum tend to want to hang on to many existing benefits, but the problem with that is that then Basic Income becomes very expensive for the state.

Several countries have been experimenting with Basic Income approaches. And in the UK the Royal Society of Arts (RSA) has been pushing the idea into public discussion – in a version that wants to hang onto other benefits.

'A Basic Income is an unconditional payment to each individual (ie it is not based on household). It is a building block for security and is designed to support the individual as they work, care (or are cared for), set up a business, or learn...

Our model works as follows:

- Payments are made to every citizen on a universal basis... Prisoners would not receive it
- The weekly amount that any working-age person receives is a 'basic' amount. In other words, if they are fit and able to work they would have a very strong incentive to do so
- All recipients over 18 would be required to be on the electoral roll, thereby reinforcing citizenship.'

They go on to propose some numbers, and it gets complicated – much more so than Milton Friedman's basic idea. But the point is: the idea is current, and it would help deal with the financial implications of lowering the number of available jobs.[13]

Once the number of people with jobs starts to go down, governments will face all kinds of problems. One is social unrest – people don't like to be out of work! But another is an economic problem – demand. If people can't buy things, the economy is on the skids. Welfare benefits aren't all about charity. One reason governments make what economists call 'transfers' to the unemployed and poor is so that they can buy stuff and keep the economy going.

So if jobs start disappearing and the idea of 'full employment' at around 90-95 per cent begins to decline, something like Basic Income is likely to result.

But the other problem, the problem of just not having a job to go to, is what to do with all the free time! This is actually a much bigger problem than it seems. Since the dawn of the modern world people have had work. This was true both before and after the Industrial Revolution. And in earlier times it was true when most people worked the land. Since life expectancy in earlier days was a lot lower than it is now, it was rather unusual to have someone still alive when he or she was too old to be able to work in the fields. The shift in life expectancy has been radical. Around 1900 it was still something like 50. Now most of us can expect to live into our 80s, at least.

The point is: it's been a new experience for people other than the very wealthy to have time on their hands when they don't need to be working...

HEAVEN OR HELL?

If Robots end up replacing workers with technology and no-one needs to work any more, is that a vision of heaven on earth – or hell? Do we all share in the wealth, and so wake up as members of a 'new leisured class' (as Keynes called it)? Or do we all wake up poor and in the situation of people today whom we call long-term unemployed?

At the moment, as we know, there are basically two groups of people who (before retirement) don't have work to do when they get up in the morning: the unemployed poor, and the very rich. How shall we handle the question of the distribution of income and wealth?

But let's assume the world solves the problem of how we manage the wealth that machines create in a manner that is fair. So, let's assume we all wake up 'rich', comparatively speaking – as economically secure as if we had been employed. The haunting question that remains is: What are we going to do with ourselves?

Keynes put this puzzle neatly in his essay on the future and technology: 'To those who sweat for their daily bread leisure is a longed-for sweet – until they get it.' In other words, it does not take long for people to switch from wishing the workday would end to being bored! The question is: what do we do with our newfound free time?

It's interesting that many wealthy people, who have no financial need to work for a living, in fact do find real jobs. Some set up foundations, such as the Gates Foundation – definitely a full-time job for Bill and Melinda Gates – and make a career out of giving away their money. Others keep themselves busy by sitting on the boards of orchestras, museums or other charities. They make themselves busy doing good works, or at least in social and cultural activities that keep them occupied. Because it is not an easy thing to have nothing to do.

If Robots bring about a workless world, the consequences for the human race are huge. As philosopher Bertrand Russell said many years ago, looking ahead: 'We shall have to change some of the fundamental assumptions upon which the world has been run ever since civilisation began.'

And we do not need to assume a full-scale workless world to need to address some of these questions. Because there's good reason to believe that the disruption ahead as machines muscle into the job market will pitch many people into 'workless worlds' of their own, whether short- or long-term.

As Robots narrow the scope for human jobs in coming decades, we should expect rising unemployment and under-employment.

We've already pointed out that many of those laid off in midlife from traditional heavy industries a generation ago never worked again. As Robots narrow the scope for human jobs in coming decades, we should expect rising unemployment and under-employment. And, of course, more people giving up on looking – which means they disappear from the statistics, as they give up hope of ever getting jobs again.

QUESTIONS TO THINK ABOUT

How likely is it that in our lifetimes we will begin to run out of jobs? Do you think paying everyone a kind of pension – what's being called Universal Income' – whether they work or not – is a good idea?

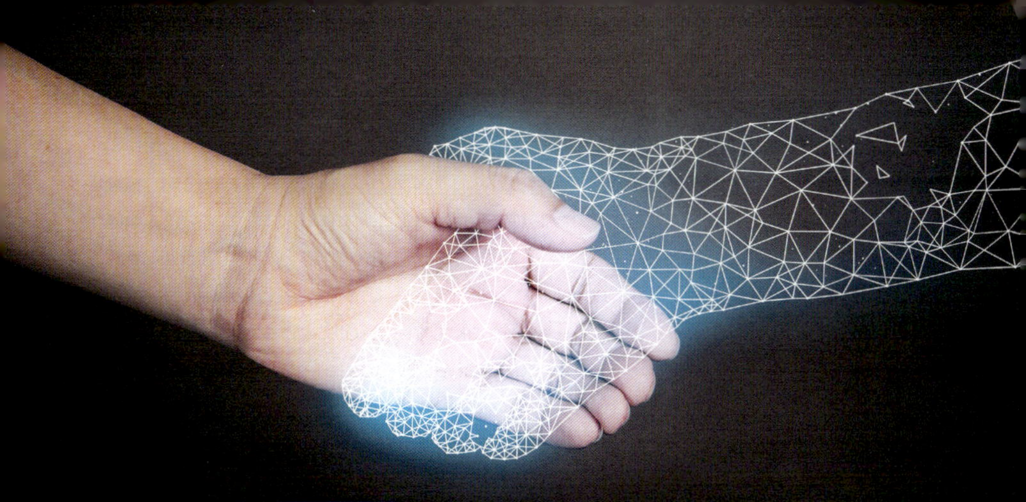

CHAPTER THIRTEEN

WHO IS MY NEIGHBOUR?

AS ROBOTS BECOME MORE LIKE US

'When we built Roomba, we explicitly designed it to not have a face. We didn't want to think it was cute; we wanted people to take it seriously, so we gave it more of an industrial look. People personified their Roomba anyway. Over 80 per cent of people name their robot. We do nothing to encourage people to do that, but they do it anyway.' Colin Angle

Back in Chapter 5 we took a look at other intelligent creatures. Angels. Animals and birds. Even, perhaps, extra-terrestrials. The reason was to put the rise of the Robots in context. While Christians believe that we humans are unique, we don't believe that we are the only intelligent creatures around. We don't believe that only we 'matter'. We don't believe that everything non-human is just stuff.

Perhaps this is a dangerous way to approach our subject. Maybe we should have kept the distinction simple: there are two categories – there are humans, and there are things. Yet that really won't work. Animals certainly aren't things, and neither are angels. And if there are extra-terrestrials, they aren't either. The universe is a little more complicated than we might prefer.

It's also a little complicated when we look at how we already relate to non-humans. We've been reminded that many animals are a lot more intelligent than we used to think. But it isn't just that. We have animals as pets. And not just dogs, many species of which are pretty smart, but cats, who certainly aren't quite as smart, and rabbits – arguably not smart at all! Many people have complicated relationships with their pets. They go to great trouble to take good care of them. They miss them when they are away. And when pets die, they grieve – sometimes just as much as when people die. We often still think of our pets, fondly and emotionally, years and years after they are gone.

And children with their toys. They can have strong emotional bonds to dolls and teddies and 'imaginary friends'. Many of us remember those bonds many years later as adults, and we can find them a little embarrassing to think about.

I think it's inevitable that we are going to have 'relationships' of some kind with Robots.

I think it's inevitable that we are going to have 'relationships' of some kind with Robots, and our relationships with pets and toys give us clues as to why, and perhaps as to how to manage things. Because there will be Robots designed as 'pets' – the next generation of Tamagotchis and Hello Barbies, designed for children but also for adults, designed to offer the kind of companionship that presently we find in animals. But, of course, designed to offer more, because they will not be 'dumb beasts'. They may be very chatty beasts! Depending how they are designed, and which ones we order, and how we decide we wish them to function. They will be as chatty and as smart and as emotionally supportive as we choose.

One of the odd things about thinking like this is that, obviously, we will know that these are Robots! They are 'just computers'. They are programmed to be nice to us. Won't that mean we don't take them seriously?

I say it's odd, because what about your dog? This may offend some dog owners and dog lovers, but certain breeds of dog – Labrador retrievers, for example – are sweet, loving, and quite intelligent animals. And while they were originally bred to be gun dogs and help hunters retrieve their game, they like nothing better than to sit around the house, play with children, romp around the garden, eat and sleep. If we treat them kindly, their behaviour is pretty much guaranteed. And yet we appreciate their friendship, we speak of their loyalty, and we never forget them.

When you read about children and old people playing with cuddly Robots – the Japanese seem especially interested in using Robot toys like that – it's obvious that something very similar is happening.

BEHAVING THEMSELVES

The idea of having Robots in our homes – especially the humanoid kind who can walk around the place – is a scary prospect. And not just in our homes. In our offices, and shops, and out in the street and the park.

The writer Isaac Asimov, whose stories have played a large part in preparing us for the coming of the Robots and inspiring many of the technology experts involved, was very much aware of this problem. He came up with a set of Robot rules, and gave them the imposing title the Three Laws of Robotics, though he actually added a fourth later. The context is of course sci-fi. They are taken from the *Handbook of Robotics, 5th edition* – published in the year 2058!

1. A robot may not injure a human being or, through inaction, allow a human being to come to harm.

2. A robot must obey the orders given it by human beings except where such orders would conflict with the First Law.

3. A robot must protect its own existence as long as such protection does not conflict with the First or Second Laws.

U.S. NAVY MQ-4C TRITON SURVEILLANCE DRONE

The fourth, which he called Law Zero and placed first, adds this: '0. A robot may not harm humanity, or, by inaction, allow humanity to come to harm.'

These may seem obvious enough applications of the so-called Golden Rule, 'do unto others as you would have them do unto you' from Matthew 7:12. But they have been very influential both in sci-fi thinking (they form a theme in Asimov's own books) and in generating the notion that Robots are a good thing and that we should not be scared of them.

Of course, some Robots are already designed for war – including the drones that fly around over dangerous parts of the world and, when a far-away human presses the trigger, fire missiles at terrorists. There's a big debate about whether they should be free to press it themselves. Will the battles of tomorrow be fought by Robots – on both sides?

In a recent press statement the US Defence Department actually made the point that on the battlefield Robots would be more reliable – and less inclined to break the rules, as human soldiers often do. They could be programmed, the spokesperson said, to obey the Geneva Conventions – the international set of principles that are supposed to govern how we fight wars. Some people now want to take this a stage further – so that there would be rules governing how Robots treat each other.

One writer in the prestigious journal Scientific American, Hutan Ashrafian, has called for another Law to be added to Asimov's list: *all robots endowed with comparable human reason and conscience should act towards one another in a spirit of brotherhood and sisterhood.*[14]

'Robot ethics' has become a new area of study. Which may be a very good idea. But it also has the effect of blurring the lines between Us and Them – and implies (as Hutan Ashrafian explicitly states) that Robots themselves have 'rights', or will do once they become more advanced.

Back in the nineteenth century when Christian leaders like William Wilberforce and the Earl of Shaftesbury were campaigning for causes like ending slavery and child labour, they were also interested in the welfare of animals. In fact, Wilberforce was one of the founders of the RSPCA (Royal Society for the Prevention of Cruelty to Animals). They saw the connection between how we treat animals and how we treat children. They were genuinely concerned about the suffering of animals, but they could also see how the kind of callousness and indifference to suffering that led to the sending of children to work down coal mines and to others living on the streets also led to the abuse of animals. Society needed to be reformed in both respects.

How does this relate to how we treat Robots? In a recent YouTube video, a four-legged Robot that looks a bit like a large dog or small horse is shown walking around a building – and being pushed over. The point is to demonstrate that it can get up again. But some people found it offensive. When we see something that looks like a dog being pushed over it affects us. We are emotional beings. How shall we cope with machines that look a lot more like animals – and like people too?

That's why watching Robin Williams in *Bicentennial Man* is, for many, a rather disturbing experience. We know he is an 'it' – a mere fancy PC on legs. But we also know – because we have sympathetic imaginations at work – that this 'it' looks and behaves so much like a 'he' that maybe it has become a he. Or is also a he. Or should, at least, be treated like a he – even if it's not, or if it's ambiguous, or if we plain can't make up our minds. When in doubt, should we extend the 'human' treatment – and be 'humane'? Of course, that is the deliberate, disturbing message of the film.

WALKS LIKE A DUCK?

There's a well-known saying, 'if it looks like a duck, swims like a duck, and quacks like a duck, then it probably is a duck.' Once we get sufficiently advanced Robots that we can chat and work with, pretty much as we do with people, what are we going to think about them?

THE TURING TEST

Back in 1950, famous British computer expert Alan Turing proposed a test to see if machines could think. It's interesting, looking back, to remember that there was tremendous confidence among scientists at the time that rapid progress would be made in the field of Artificial Intelligence. Since as we've already noted our computing power since then has increased by more than one trillion times (that's 1,000,000,000,000) that confidence looks a little shaky! And we may need to be wary of the similar confidence being projected by people like Ray Kurzweil as they forecast rapid developments today, nearly 70 years later.

Turing recently became even more famous as a result of the film *The Imitation Game*, that covered his work at Bletchley Park, the top-secret centre for Britain's codebreakers during the Second World War.

Critics of the film have pointed out that a lot of it is fiction, but what's interesting for us is its title. In the article that Turing wrote proposing a test for 'thinking' machines he said this: 'Are there imaginable digital computers which would do well in the *imitation game*?' While Turing's original idea was a little more complicated, the 'Turing Test' boils down to whether a computer can fool a human into thinking it is actually human – whether it can 'imitate' sufficiently well. As we have already seen, by winning the TV game show Jeopardy and defeating the top players of both chess and the Asian game Go, computers are now able to challenge both human general knowledge and intellectual skill.

STATUE OF ALAN TURING AT BLETCHLEY PARK

But the contexts of course are abstract. The question is: will the day come when you have someone living next door to you, or working alongside you in the office, who is actually – picking up the categories that we noted from Spielberg's movie *AI: Artificial Intelligence* – 'mecha' rather than 'orga'? Perhaps that would be the Turing Test for the twenty-first century.

QUESTIONS TO THINK ABOUT

What do you think God thinks about Robots? If they get really smart, might he want us to see them as the equivalent of people?

CHAPTER FOURTEEN

HARD QUESTIONS FOR THE HUMAN RACE

'I am not a robot. I have a heart and I bleed.' Serena Williams

'No one knows when a robot will approach human intelligence, but I suspect it will be late in the twentieth century. Will they be dangerous? Possibly. So I suggest we put a chip in their brain to shut them off if they have murderous thoughts.' Michio Kaku, Physicist

This book is full of questions. There is so much that we just don't know. But we do know we need to get prepared. We know that the Robot future may not all be wonderful. But we also know that if we get the questions right we are at least on the road to the answers.

Later on we pose some questions specially for Christians. But many of the issues Robots raise are questions for everybody, whatever our faith or lack of it. They're human questions, and everyone who loves being human and loves other humans needs to face them.

1. ARE ROBOTS JUST GOING TO MAKE THE RICH RICHER AND THE POOR POORER?

Some of the enthusiasts for Robots talk of 'plenty' – we shall soon have so much wealth, so many resources, so much amazing new inventiveness, that no-one will ever be hungry or cold again. But the evidence so far is not quite so encouraging. While poverty worldwide is actually slowly going down, there's no evidence that Moore's Law – which means digital tech keeps improving faster and faster – is about to make starving Ethiopians or even Americans who are currently just making ends meet into billionaires.

It's obvious that as jobs migrate to Robots what is happening at an economic level is that capital, in other words, machines bought with money, is replacing labour as a 'factor of production'. In other words, investors are taking over from people in making things and offering services.

The traditional view is that there are basically three factors of production – labour (people), capital (money), and land. Some people would add entrepreneurship to the equation – the factor that brings the rest together and makes it all work. Either way, if people are becoming less significant, and capital (as in, Robots!) is taking their place, it's looking bad for humans. Even if some people do keep their jobs, and some new jobs are created, the trend is away from people as resources – as machines, aka Robots, replace them; and as capital, aka investors, reap the rewards.

Obviously, governments are going to get deeply involved in managing these changes, and it's likely that in different countries they will take rather different approaches.

2. IF ROBOTS DO ALL OUR WORK, WILL WE HUMANS BE ABLE TO FILL OUR TIME?

This may not seem the most obvious question. When we think of Robots taking over our first concern is about money. Who will pay you if you don't have 'a job'? But the question of what you will do with your time – what we will do, with our time – is actually more difficult. It's pretty much inevitable that governments will pay people something if they don't have jobs, and there are two basic reasons why.

For one thing, governments needs votes, and governments do not do well in elections if a lot of people are out of work. So if the unemployment numbers go up and up, and we move away from the standard (post-Second World War) assumption of around 90 per cent plus employed, governments of right or left are likely to want to be relatively generous to the jobless. Whether they regard them as 'long-term unemployed' or members of the 'new leisured class' which John Maynard Keynes predicted.

There are two fundamental elements in an economy: production, which the Robots are taking over, and demand.

The second reason is equally simple. There are two fundamental elements in an economy: production, which the Robots are taking over, and demand. Whatever the Robots are making needs someone to buy it. Robots aren't going to buy it! If a lot of people are living on the poverty line, we have a problem.

So, it's likely that people whom the Robots put out of work will be paid something, even if not a lot. The *bigger* issue is actually what they will do with their time.

We all like finishing work. And many of us get pretty bored once it's over. On a larger scale we know that communities where there aren't a lot of jobs don't do well. The unemployed don't seem very good at using all that 'free time'. Not all of us enjoy reading, gardening, volunteering for charity and spending time with our families enough to keep us occupied day in and day out. Some people just fall apart when they are out of work, and for the same reason many people find retiring very difficult.

Of course, there are some – not many! – who don't ever need to work. Either they were born with a lot of money, or they have been successful in business and made a lot of it. (Or maybe they won the

National Lottery!) It's interesting to see what many of these lucky people end up doing – they make up jobs for themselves. As we've seen earlier, they set up charities and foundations, which are often full-time jobs to run. Or they become regular volunteers, passing on their skills in the community. Wealthy people – the people who don't *need* to work for a living – usually end up working anyway. Is there something in human nature than means we need to have occupations?

3. WHO WILL DECIDE HOW ROBOTS MAKE DECISIONS?

This is not just a long-term issue: we already have self-driving cars on the roads. How do they decide what to do if they see a child run out and the choice is 'kill the child or risk killing the driver?'

Since self-driving cars, or 'autonomous vehicles' as they are more formally known, are already a fact, one core issue has focused on their decision-making. Of course, when a human driver faces a crisis situation – a cyclist wobbles, or a child runs into the road, or a dog – we just don't know how we will respond. We hope we will make the right choice, whatever 'the right choice' actually is. But our responses are emotional, we may find ourselves panicking, and we don't know all kinds of facts about the consequences of swerving, braking, and making other emergency moves – facts that a Robot driver may well know. And, of course, Robots don't panic.

The point is, exactly how a Robot driver responds in such a crisis situation will have been planned by the people who programmed it. And, behind them, will lie the policies set out by the company's executives, and their lawyers. How will Robot cars make choices? There is nothing random about what they will do.

Will they decide that the driver of the car, who after all is the customer of the Robot company, is their priority? Will they choose the child, even if there is a risk to the driver? And what if it's an older person in the road instead of a child. What if he is 35? Or she is 70? Or he or she is plainly a drunk or a vagrant? All of these facts will be programmed into the machine, and will decide what it decides. Because 'it' won't be deciding anything. The people who run the Robot company will be guided, of course, by the law. So we can expect big issues to be faced by politicians if they decide to set boundaries for these private choices.

If you read articles about self-driving cars you will see a lot of discussion about what's called the 'trolley problem'. It's an old idea in ethics that experts have written about over the years, but it's become very relevant now that there are self-driving cars on the road. The basic idea is this: a 'trolley' – a tram or a train – is coming down the track and there are five people lying on the track ahead of it. But there's another track that the train could turn off on.

You're standing there and you see a lever you can pull that would save the people by turning the trolley onto another track. However, there is someone else – just one person – wandering along on that one. Do you leave the trolley to its own devices, or pull the lever, save five people, and kill someone else?

Part of the problem that this example reveals is the new difficulties that we have now we know more than we used to. If you are programming a self-driving car, you need to have answered this question and many others like it in your head.

Is this a task for an individual programmer? A company? Should governments decide these things? Should the individual car-owner?

4. HOW WILL WE COPE WITH ROBOTS IF THEY GET TO BE VERY LIKE PEOPLE?

This is hard. I mean, if they are *really* like people, do we have to make sure we treat them like people? Or do we have to make sure that we make sure we don't? You can argue it either way. It may be the biggest challenge the human race has ever faced.

I suppose there are just these two options – unless, of course, Robots take over (see number 6). Do we receive them into the 'human'

community as if they were members of our human species? Or do we make it our core principle *not* to? However respectfully we treat these pseudo-humans, do we decide that they need to stay machines; our machines; our servants, our slaves? Never, ever, treat them as 'people like us?'

5. IF WE END UP HAVING ROBOT 'SLAVES', WHAT WOULD THAT DO TO US AS 'SLAVE-MASTERS' OF MACHINES THAT LOOK AND TALK AND THINK VERY LIKE US?

If we decide that Robots are just menial machines like vacuum cleaners to do jobs for us, what will it mean for our humanity if these machines chat with us like friends? 'Father of cybernetics' Nobert Wiener, who foresaw the internet and more back in 1930, said that the Robot economy would be just like the slave economy of America before the Civil War. We would have slaves.

If Robots look and talk like people and we treat them as 'things', what does that say about us?

The question I am asking is, If Robots look and talk like people and we treat them as 'things', what does that say about us? From one point of view it's simple: we are people, and Robots are fancy stuff. From another it's not: we are people, and Robots are so much like us that treating them like slaves really hurts our 'humanness'. I'm not suggesting an answer here, but I am definitely proposing the way we need to frame the question.

CHAPTER FIFTEEN

FREQUENTLY ASKED QUESTIONS

Questions people often ask! With a little more clarification.

1. WILL ROBOTS TAKE OUR JOBS?

Plainly, they have already started to. Robots drive the trains on London's Docklands Light Railway as they do at many airports. Robots, in the form of cash machines or ATMs, have taken over jobs from bank tellers. Robots book your tickets when you travel – there aren't many real travel agents left – the travel shops in the High Street mainly sell package holidays. And of course Robots – in the form of smart mobile phones – now take almost all our photographs. It's become much harder to earn a living as a photographer.

But the big question is whether, over time, Robot technology will create all sorts of new jobs – including people to make and maintain the Robots! – to take the place of jobs that are going out of existence. The short answer is that nobody knows. But there are smart people who think that this time it's different. New technologies in the past caused job losses but always ended up creating plenty of new jobs. Perhaps this time we shall end up with fewer jobs to go around – and more 'leisure' for more of us.

2. AS YOUNG PEOPLE MAKE CHOICES – AND AS OLDER FOLK ADVISE THEIR KIDS

What jobs are Robot-proof?

I wish there were a simple answer. One answer that is plain wrong is this: get everyone to focus on technology, or the so-called STEM subjects (science, technology, engineering, maths). It's not entirely wrong – obviously, knowing more about technology will help all of us in the Robot future. But, certainly at a basic level, the kind of jobs that rely on technology skills are going to be the easiest ones to be done by Robots. Just look at the long list of jobs that the Oxford report (see the list of readings at the end of the book) reckoned would go first to the machines.

But in practical terms, the big need is to be flexible, and learn the kind of skills – whether at school and university, or on your own – that help you be adaptable. Work on your imagination to help you be creative. Learn basic business skills so you can be self-employed or maybe start your own business. Think about the kind of jobs where people – like you – would really prefer to be dealing with a person rather than a machine. Perhaps you would be prepared to pay extra!

For example, already some cafes and bars are starting to use tablets to take our orders. Many of us would be very happy to pay a little more for human service. But it gets complicated. Already there are computers giving counselling and 'reading' people's emotions, and there's evidence some people actually prefer talking to a Robot rather than a person when it comes to private, sensitive issues. So don't assume that 'touchy-feely' jobs like psychology will be safe for people.

3. HOW LONG WILL IT BE BEFORE ROBOTS TAKE OVER?

Well, we don't know if they ever will, and it does depend on what 'take over' means! They've already taken over a lot of things, and – this is important – we have already become very dependent on them. Think how you feel when you forget your mobile. Think how you feel if it breaks down and you don't have one for a day or two. Think what your life would be like if you never had one again.

Inventor and futurist Ray Kurzweil, who is controversial but has been quite good at making other predictions, tells us that what he calls the 'Singularity' – when Robot intelligence takes off and could take charge – will come in the year 2045. But he also reckons that by 2029 we shall be able to live for ever, as medical research will keep lengthening our lives every year by a year after that date, so he is a bit of an optimist!

What may be most useful to know: anything to do with Robots and computer intelligence is speeding up all the time. Don't believe people who say 'that will never happen'. Whatever it is, if it's Robot-driven, it could, and sooner than you expect.

Don't believe people who say 'that will never happen'. Whatever it is, if it's Robot-driven, it could, and sooner than you expect.

4. SURELY ROBOTS WON'T BE ABLE TO DO THINGS LIKE WRITING NEWS, OR FICTION, OR POETRY? WON'T THESE HUMAN JOBS BE SAFE?

Sadly, no. At least, we already have Robots writing basic news stories, and they are already capable of producing fiction. It may not be world-class, but it's fiction. A Robot writing great literature may never happen, but the kind of thrillers and romances we buy at the airport to pass the time on the flight? Perhaps.

5. IF ROBOTS TAKE OUR JOBS, HOW WILL WE PAY OUR BILLS?

One reason I think leaders and governments around the world are being short-sighted in refusing to think seriously about the threat that Robots pose to jobs is exactly this. If the need for human workers goes down, are we all going to be able to retire earlier, or does it mean we shall spend half our lives on the dole? Plainly, if Robots are doing our jobs then the wealth that we have been creating in the past will still be created by the smart machines that have made us redundant. The issue is where the wealth goes, when machinery does what humans used to. It's obvious that governments are going to have re-arrange things so that we don't all die of starvation, and they will probably do it by something like taxing businesses that use Robots to do what humans used to do.

6. WILL WE REALLY GET SELF-DRIVING CARS? IT JUST SEEMS SO UNLIKELY, AND DANGEROUS.

In fact, self-driving cars – or 'autonomous vehicles' as they are called in the trade – are one of the dead certs of the Robot revolution. We already have them, and they work. They have already proved themselves safer than human drivers.

One fact that people often ignore when they ask this question is: we humans are not good drivers at all. We are very unsafe. We spend around half the time day-dreaming, and miraculously our eyes and brains keep just focused enough on the road to stop us crashing – while we work out what to have for dinner or re-think that difficult conversation at work or listen to music, or just feel dangerously tired. That's why there are an awful lot of accidents. Britain actually has some of the safest roads in the world, but even so in 2015 there were 1700 people killed, and 22,000 seriously injured, in road accidents. In the US, the number of deaths was 38,300, a far higher rate. The point is, to be safer, the Robot cars just have to be better than bad human drivers. They don't listen to the radio, think about the shopping, and they certainly don't send texts, get tired or chat on the phone. Of course, with a Robot driving we shall be able to do all these things quite safely!

So far the Robot cars have been tested on easy roads in easy conditions. Navigating a single-track road in the Scottish Highlands in the middle of winter will prove a lot harder. But we're on our way.

7. WHAT ABOUT PRIVACY, IF ROBOTS KNOW EVERYTHING ABOUT US?

Good question! But since Robots already know pretty much everything about us, it's not really a new issue. Just more of the same. And there's no doubt that as Robots move in on fresh areas of our lives part of the attraction will be the kind of service/privacy trade-off we have been getting used to for the past 20 years. Facebook is free. Google is free. But not exactly. Facebook and Google offer us free services in exchange for sucking up and analysing huge amounts of our info, and using it to target us with ads.

There are special privacy problems associated with the self-driving cars, of course. Will they report us for speeding – assuming we are able to tell the car if we are in a hurry? Will our insurance companies see all the details of our driving, Robot and personal, and be able to change our rates?

8. AREN'T WE MAKING A FUSS OVER NOTHING? THESE ARE JUST MACHINES.

It would be nice if you were right! I was at a conference recently when a computer science professor said just that. From his point of view these are just like PCs, even if they are dressed up in other shapes and sizes and roles.

The problems aren't just with the Robots, they are with us. What I mean is that we humans naturally tend to treat other things as if they were like people. We do this with our pets. Children do it with their toys. And when we watch the spooky films we have been talking about, we realise how easy it is for us to get wrapped up in some kind of 'relationship' with something that we know to be a something – even if sometimes it seems like a someone.

Of course, children are a special case, as they have always played imaginatively with toys and given them personalities and had conversations with them – these are key parts of their development. Will their having toys like Hello Barbie hinder that development? Perhaps, though perhaps not. Older people may be a special case too – if they are losing their memories and perhaps their grip on reality. Should we have them cared for by machines that talk like people – and maybe look like people?

The talk question is important, and for all of us. One of the most difficult and also exciting developments in Robot technology has been 'natural language' – making machines that can listen to us talking (rather than just receiving commands from us in computer code), and then talking back to us. Because this is how we communicate with our fellow humans, it's hard to see how we will avoid confusing machines and humans if they listen and talk in just the same way.

OK, there will be some people – computer science professors, anyway – who won't have a problem treating machines as machines. But the rest of us? I'm not so sure, and I don't think you should be either.

9. HOW CAN WE PROTECT OUR CHILDREN FROM THESE INFLUENCES?

It's complicated, not least as our children may also be those who best understand what's going on. And they certainly are not likely to want to listen to us 'telling' them not to do this or that. The best approach is to collaborate with them – so we can learn from their experience, and help them to ask questions about it.

The need to ask questions has always been one of the keys to successful parenting. The best example may be the old-time issue of watching TV. Back when families had one TV – back in the 1950s or 1970s – everyone sat around it. Some families rationed children's watching. In our family the children were each allowed to pick one hour a week on evenings before school days – with more freedom at the weekend. But more important than how much they watch is how we discuss *what* they watch. That's true of the ads, but it's also true of the programmes. What's going on here? What's the message? How can we avoid having our minds shaped by what we are watching – and learn from it?

That kind of parent-child discussion – much more difficult now that we all have separate screens and children's and teenagers' viewing is harder for parents to monitor – remains vital. And the point here is: it needs to guide our approach to these emerging technology issues and the many kinds of Robots they will encounter.

The key is to give children a questioning mind.

Whether 'protect' is the right word or not, the key is to give children a questioning mind. They need to learn to challenge, to dispute, to be cynical. To ask the hard questions – and the human questions. These are tools that will enable them to manage situations that today we can't begin to imagine, as the technologies ramp up and intersect with us and them and their own children into the future.

10. WHAT DOES GOD THINK ABOUT ALL THIS?

When something takes us by surprise it's hard to avoid thinking it must be taking God by surprise too! But of course that's ridiculous. God may be saddened by what humans get up to, but He's God so He surely is never surprised by it.

We tend to think of Him back in Bible times, when there was a lot less technology around. Or we think of Him back in our grandparents' and parents' day. Or, even, in the days when we were younger. But God is out there in the future, and unless He decides to intervene and bring our world to an end he may be planning for humans and our technologies to be around for another thousand years. Or another million years. Or perhaps another billion years. That boggles our minds, and it should. My point is: we have to take a long-term view, and not assume that anything that comes from human technologies – whether the atomic bomb or television or the internet or Robots – is 'the end of the world'.

These are all challenges for humans like us to manage, and they all have aspects that are very scary.

But we believe in a God who made space and time. He made them and built into them all the technology possibilities humans have ever devised or will ever devise. And He commanded us, his human creatures, to have dominion – to subdue the earth. We don't always do it right, and we are sometimes very puzzled as to what doing it right would be. But the latest technologies aren't a surprise to God, and I don't think they should be to us.

CHAPTER SIXTEEN

ISSUES FOR THOUGHT, DISCUSSION, AND PRAYER

There are no simple answers to any of the questions raised in this book. We have highlighted some at the end of each chapter. More talking points here...

1. GOD GAVE US 'DOMINION' OVER THE EARTH. ARE THERE LIMITS TO THAT DOMINION?

We know there are limits of various kinds. Plainly, God will hold us accountable for how we have used our God-given abilities to make tools, Robots included, and use them. We must respect our fellow humans, made in his image, and we know that there are also implications for how we deal with ourselves. We can't take our own lives any more than the lives of others! And there are limits to how much we can manipulate human life. Are there limits to how we build and empower Robots to do things for us? Is it too much of a risk to go much further? Can we trust God to save the human race from the Robocopalypse? Or is that a wrong way to look at our trust in Him?

2. SOME OF THE MOST INTERESTING, AND USEFUL, JOBS FOR ROBOTS INVOLVE THE CARE OF THE ELDERLY AND ALSO THE CARE OF CHILDREN.

Children and the elderly need a lot of care. It's expensive, time-consuming, and for many people – especially with the elderly – a job without much satisfaction. Already there are experiments going on, especially in Japan, using Robot toys to chat with elderly people and keep them company, and to calm children. There are Robots that undertake basic nursing tasks, and can enable older people to live alone rather than in hospitals and old people's homes. These uses of Robots are just in their infancy – this will likely become a huge business and save families and governments a lot of money. Does it worry us that at the start and end of life we are looking at ways of giving people machine care rather than human care? There's a special problem in that both small children and older people with dementia may confuse Robots with people. Is that a problem? Does our Christian belief that every human being bears the image of God make a difference?

Does it worry us that at the start and end of life we are looking at ways of giving people machine care rather than human care?

3. IF IT'S POSSIBLE THAT IN THE NEXT TWENTY YEARS A LOT OF JOBS ARE GOING TO GO TO ROBOTS, WE NEED TO THINK AHEAD.

What would it be like if everyone could retire early? Or if we could only work part-time? Or if while some people carried on with full-time jobs pretty much as they do now, the number of those without jobs just kept going up? There are two kinds of problems here. One is financial – and we would expect governments to find ways to make sure that people could pay their bills. But the other problem, though less obvious, may prove much bigger. What will we do with our time? As we have seen, many people who are retired with pensions, and the very wealthy, don't need to work. We know that people who are retired often find it hard to fill their time. The Church has quite a bit of experience using retired people to help with pastoral and other activities, as well as ministering to the unemployed and helping them find ways to use their skills – and find meaning. How would it cope if there were a steadily growing number, either early retired when still fit and full of energy, or a growing number of long-term unemployed, or both?

4. IS IT GOOD FOR US TO BE SO DEPENDENT ON OUR MOBILES AND OTHER ROBOT DEVICES?

Would it be helpful for Christians to be among the few who, while welcoming the amazing things technology can do, are also able to distance themselves from it?

Little things help, like in our families or perhaps Church-wide, agreeing not to sleep next to our mobiles, to read a book rather than check a device last thing at night, to be careful never to have mobiles at the dining table?

And what about a technology Sabbath, which some families practice? Parents often limit 'screen-time' for their kids, but this goes a lot further – as it's not just children who will benefit! Whether it's Sunday or Saturday – one day when we as individuals, or families, or Church congregations, agree we will turn off all our devices and live like we used to – back in what seems like the Middle Ages but was actually only 1980?

It's important to see efforts like this as 'pro' technology, not 'anti'. It gives us a much better grasp of what technology is all about if we know what it's like to live without it. And since in ten years' time technology, in the form of Robots of one kind and another, will probably be ten times as significant as it is now, and this may be one of the best ways we can begin to prepare.

CHAPTER SEVENTEEN

WHERE DO WE GO FROM HERE?

What's the biggest problem we are likely to face as the Robots get smarter and there are more of them and we have to deal with them day by day? It's getting confused. Confused about what makes us humans special and different, as we keep coming across machines that are as good as we are at doing something, whatever that something is.

What we shall need to keep reminding ourselves is that humans are humans and machines are machines. That's not difficult if the machine is an old-fashioned vacuum cleaner, especially if it's one of those that gives off a burning smell and gets hair tangled round the roller and is messy to empty. It's a bit more difficult if it's a Roomba that cleans the floor on its own – though at the moment we still need to empty it and put it on charge overnight.

But it's going to be a whole lot more difficult when we're dealing with a human-like servant like Andrew in *Bicentennial Man*. That's true even with the Andrew we meet at the beginning of the film, who is a clunky Robot-looking Robot. As the film goes on and Andrew keeps getting upgraded it only gets harder.

Remember back in Chapter 3 we quoted the famous parson and poet George Herbert? His hymn *Teach me my God and King, In all things Thee to see* gives us a wonderful summary of Christian living in the middle of ordinary daily life. The key to the hymn – and the key to living as a Christian believer – is the little phrase 'For Thy sake'. If we want to understand how to live as Christians in the high-tech Robot world of tomorrow, maybe we should start by going back to basics.

GEORGE HERBERT'S HYMN

Let's read the hymn through. If you know the tune, you can sing it to yourself! The hymn-books usually miss out the second verse, but I'm quoting it here as it's worth thinking about.

1. Teach me, my God and King,
   In all things Thee to see,
   And what I do in anything
   To do it as for Thee.

2. Not rudely, as a beast,
   To run into an action;
   But still to make Thee prepossest,
   And give it his perfection.

3. A man that looks on glass,
   On it may stay his eye;
   Or if he pleaseth, through it pass,
   And then the heav'n espy.

4. All may of Thee partake:
   Nothing can be so mean,
   Which with his tincture—'for Thy sake'—
   Will not grow bright and clean.

5. A servant with this clause
Makes drudgery divine:
Who sweeps a room as for Thy laws,
Makes that and th' action fine.

6. This is the famous stone
That turneth all to gold;
For that which God doth touch and own
Cannot for less be told.

Herbert wrote this poem in 1633, and he certainly wasn't thinking about Robots. But what he *was* thinking about was the vast difference it makes if we do everything we do 'for Thy sake'.

He gave his poem a title: *The Elixir*. To understand why, you need to understand a little about something that sounds very strange in the modern world of science and technology: *Alchemy*. Alchemy was the idea that, if you experimented with chemicals for long enough, you could turn things into gold. For hundreds and hundreds of years many of the world's smartest people spent their time trying and trying again to do just that. In 1633 everyone knew what he was talking about and, like the wonderful preacher that he was, Herbert used ideas that everyone understood to make his Christian points.

God will help you turn the ordinary things in your life – into gold!

And the idea fits very well. People wanted to turn ordinary things into gold. Here, he says, God will help you turn the ordinary things in your life – into gold! We have this special medicine – this 'elixir' – and it will do the trick. What is it? It's to live for Jesus. It makes all the difference.

So back to the hymn.

Verse 1 is simple enough: It's a prayer to God to teach us to see Him everywhere we go.

Verse 2 looks a bit odd, but just means 'Don't run unthinkingly – 'like a beast' – into everything you do; think about it. And do your very, very best'. Give it your 'perfection'.

Verse 3 uses an analogy – a window. You can either look *through* the window, or you can look *at* the window. If you're a window-cleaner, or maybe a glazier, you need to look *at* the window. But what are windows for? They're for looking *through*. What happens when you look through the window – you see heaven! So learn to look *through* the tasks you have to do today – and see heaven in everything you do. What a wonderful thought.

Verse 4 underlines the point with another illustration. There's absolutely nothing too 'mean', if you use the special 'tincture' (ointment) – 'for Thy sake'. It will scrub and shine every single thing in your day. Just do it for God.

Verse 5 gives an example of just about the most tedious and 'mean' human task, sweeping a room. Remember, this was back before vacuum cleaners – even the old kind that give out burning smells and get tangled hair in their rollers. All you had back in 1633 was a broom. And if you were a servant, you could spend a lot of your long, tedious day sweeping with it. So Herbert says: 'All may of Thee partake'. Everyone's included! Even servants who spend their days sweeping. Because you can sweep for God. Doing it for Him will be so magical that it will make even 'drudgery' divine.

Verse 6 sums it all up. And here he uses another of the favourite words of the Alchemists. They had this idea that if they kept looking they would find what they called the 'philosopher's stone', the magical substance that would turn everything into gold – and some of them believed it would give you everlasting life too! You've probably heard of the 'philosopher's stone' thanks to Harry Potter.

Well, here it is.

This is the famous stone,
That turneth all to gold;
For that which God doth touch and own
Cannot for less be told.

Aside from showing us what a brilliant communicator George Herbert was back in 1633, this little hymn-poem provides us with a powerful message nearly 500 years later. Doing things for God makes all the difference. It makes drudgery divine. Nothing, and no-one, is too 'mean' or insignificant for Him. Turn everything to gold and live forever – by saying 'For Thy sake'!

FIVE CLUES TO BEARING GOD'S IMAGE IN THE ROBOT WORLD

We began by asking what it is that makes us humans unique. According to the Bible, it is that we are made in God's image. But it's a rather mysterious phrase. Unpacking it will help us stay focused as human beings in a world in which machines are going to have a bigger and bigger role.

Let's go back to those five key aspects of the 'image of God' that we outlined back in Chapter 3.

First there's a focus on our brains – our ability to reason. But not just reason in the sense of argue. The human brain is also where we make decision, choices, and have free will. Of course, machines can make decisions – but only decisions based on the rules the people who made them programmed them with.

Humans make choices, which makes us accountable – we humans are responsible people. We can do good, and we can do wrong. So the second component: We can regret, we can repent, we can apologise. We can forgive – and we can be forgiven. These are all 'human' things. Of course, that doesn't mean only humans have all of these things. Anyone with a dog knows dogs can feel guilty! You arrive home and it slinks away into the corner, and you know she's done something wrong long before you discover one of your cushions has been chewed to pieces.

Thirdly, our human brains go well beyond 'rational' things like arguments and choices, and even moral things like right and wrong and guilt. We humans are also creative – we can also reach beyond ourselves in art and poetry and music. From our children's first drawings to Michelangelo's astonishing painting on the ceiling of the Sistine Chapel we humans can use our brains in all kinds of ways. And this package of 'human' qualities that flows from our brains reflects what God himself is like – he speaks and argues, but he also creates.

Fourth: we are called to live on this earth on God's behalf. We 'image' him as those he has placed in charge, his stewards. He has given us dominion over the Earth. This is all part of our reflecting his rule over the universe, and of course has led directly to the reason for this book.

Our dominion over the Earth led to our developing science and technology. And the top of the technology tree is digital technology. And in the very top branches of the digital revolution is Artificial Intelligence and the development of Robots. That's why this discussion is so important. We have been obeying God's instructions by working on science and technology and coming up with these amazing inventions. The question is, what do we do with them? If we really are 'stewards' – if the stuff we use, like the silicon that we use to make computer circuits, actually belongs to God, like the brains and hands and eyes He has given us to work with that stuff – how are we to use these gifts?

The final element we discussed was this: we are creatures who are made for relationships. Both human relationships, and also a relationship with God himself. We aren't just brains, we aren't just rulers, we are women and men and boys and girls in families and communities, with friends and relations. Not everyone is very outgoing – some and shy, and find making new friends hard. But it's often the shy people who value most the friends they do have – and their relatives.

One of the most interesting things about the Bible is that it's mostly stories. The stories of people, the story of Israel, the story of the ministry of Jesus, and all those parables, the story of the missionary journeys of the Apostle Paul and the expansion of the early Church.

Of course, it's much more than stories. It's poems – the Psalms especially, 150 poems stuck right in the middle of the Bible. It's Proverbs – all those little sayings that catch hold of the details of our lives and offer wisdom in compact nuggets for what someone called the 'small change' of life. It's Letters, some of them focused on complicated arguments – but they are Letters, located in the lives of their readers and of their writer. And that's true whether we are thinking of the real-life Letters of the Apostle Paul, or the more formal Letters to the Seven Churches that open the Book of Revelation.

What's so interesting is that it's when we read stories and tell stories that all these human pieces fall into place: the whole lives of people, their brains, their decisions, their responsibility, their relationships. Because while these are all different aspects to the 'image of God' that makes us human, they can't quite be separated from each other.

It's also why what the press calls 'human interest' stories are so powerful. Nothing interests us more than people. It's why we love catching up with friends, and why (even though we may be embarrassed to admit it) we are fascinated by the latest gossip about the lives of celebrities. It's why stories – Bible stories and all stories – so quickly grab our imaginations. They get their power from the fact they are about people, and people are images of God.

So most of us love curling up with a book. All of us love to sit and watch films, and binge-watch series – which is why Netflix and other on-demand services are proving so immensely popular. We are fascinated by being drawn into the lives of other people, both those who are real and those who live only in fiction. And while our love of fictional characters may seem odd, it's pretty much universal. Any strongly-drawn character from a favourite book or film will live on in your life, floating around in your mind together with all the real people and situations you remember – perhaps for the rest of your days. This love of people and their stories is probably the single most 'human' thing about us. Just think about the enduring love we have for the novels and characters of Jane Austen.

God has given us brains that can reason, make moral choices, and also be creative.

While some Christians thinking about the 'image of God' have emphasised one or two of these five strands more than the others, all five go very well together. God has given us brains that can reason, make moral choices, and also be creative. He has placed us here to rule the earth on His behalf. And He has made us creatures who need relationships, with each other and with Him. These are all strands in our being made in his image. The people whom we most admire, or wish to be, tend to be those in whom all five are obvious and strong. And when someone seems to lack one or more, either because of handicap or accident or for some other reason, we feel sorry for them or uneasy about being with them, or uncertain what to make of them.

Of course, all these qualities – components in the image of God – that are true of us can be copied. People write stories about Robots as well as about people, and while some of the Robots are just plain 'machines' from start to finish like the out-of-control super-computer HAL in Arthur C. Clarke's *2001: A Space Odyssey*, there are others a lot more like us. That isn't just true of Andrew in *Bicentennial Man*, it's true of cute little R2D2 in *Star Wars*. But of course the reason they appeal to us is that they have stories like our stories – it's the human side, or the fake human side (or in the case of R2D2, maybe the fake pet side), that grips us.

One reason we find human stories so fascinating is that, for all the differences between us as individuals and the stories we are reading, we feel kinship. They are 'people like us'. This is especially true of the celebrity human interest story. It may seem odd, but the reason we find it so fascinating is that the really interesting stories aren't about ocean-going yachts or ridiculous salaries or gold-plated aeroplanes, but relationships. Who doesn't know that Angelina and Brad have split up? Or that Mr. Darcy and Miss Bennet finally got together? They have relationship issues like the rest of us.

It's the things 'just like us' that appeal to us in people who are also very different. And when it comes to Robots, the most difficult thing of all will be coping with ways in which they also are like us. With every advance in the technology, our ability to copy, imitate, or to use a stronger word *fake*, human characteristics will get more potent.

In the book I wrote ten years back with Christian leader and advocate for the disabled Joni Eareckson Tada, *How to be a Christian in a Brave New World*, we came up with a little phrase that captures the many challenges posed by science and technology for a Christian view of human life in the twenty-first century. *Taking Life, Making Life, and Faking Life*. An issue like euthanasia in which some people argue we should take people's lives is complicated, though many of us believe it is always wrong. The *making* of human life using test-tube baby techniques – or maybe using cloning! – has caused a lot of controversy, and as we become able to design our babies it will get even more so. What we are dealing with here in the Robot world is *faking* life. Does that make it wrong? Perhaps, perhaps not. But it makes it complicated.

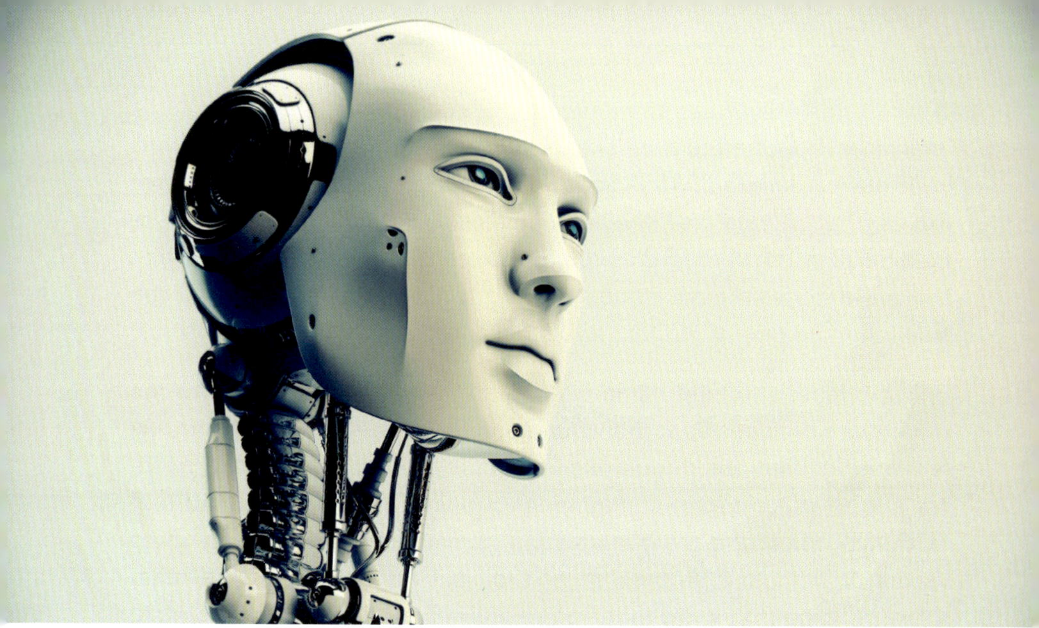

Back in Chapters 7 and 8 we looked at the wide range of technologies already out there and helping us live our lives – from booking travel tickets to giving us driving directions to vacuuming our living rooms. Things start getting complicated when they get blurred.

One of the most memorable things in Stephen Spielberg's film *AI: Artificial Intelligence* is the clear distinction drawn between the people and the machines in that society. As we saw, the people are 'orga' (from the word organic), the machines are called 'mecha' (they are mechanical). Of course, the theme of the film is the problem of maintaining this distinction, when mecha creatures become very like orga. The fair where mecha are destroyed to entertain the crowds is a horrible spectacle, because of course the cute human-like Robot boy almost becomes a victim. No-one said living in a time of Robots was going to be easy.

And yet the mecha versus orga distinction is vital. We people are orga, made in the image of God. These things, however smart machines become, and however much they are made to look like people, are still things – mecha. But does this mean we can just do what we like with them?

From one angle, it looks like the answer should be Yes. If I am done with my old laptop, the only issue is how I dispose of it in a way that will not cause environmental damage, and maybe enable parts of it to be recycled. When you are done with the annoying voice that your bank or credit card company has decided to use to help its customers, you can just put the phone down. But what about 'David', the boy-like mecha fake human, in *AI: Artificial Intelligence*?

The answer probably is this. Just because a creature is mecha does not mean we should feel free to abuse it. There are parallels in the way we treat animals (although, yes, there are also differences). As we've seen, nineteenth-century campaigners against slavery and child labour like Wilberforce also campaigned for animal welfare. One reason was that they thought Christians should be kind to animals. But another reason was that they knew that people who were cruel to animals were more likely to be cruel to children. This second point becomes extremely important when it comes to Robots that begin to resemble people.

Remember, even the sex robot campaigner David Levy was embarrassed when a journalist asked him about child-looking sex robots for paedophiles. Just because a creature is a thing, a fake, mecha, does not mean we should feel OK about treating it as if there was nothing moral going on.

Of course, as that example shows, the moral issues of the situation extend to the real people involved. The fair scene in *AI: Artificial Intelligence* illustrates the same principle. Are there limits to what we should do, as individuals and maybe as a society, to a creature designed to look and behave very like one of us? Not because mecha has miraculously become orga, but because of those of us who are orga. We so value human life and human dignity – in Christian terms, the image of God is so important to us – that we will want to protect even a fake copy of that image.

> *What if we do end up with Robots as smart as we are, or smarter?*

But what if we do end up with Robots as smart as we are, or smarter? If some of the tech gurus are right, this could happen in less than 30 years. Others think it may never happen. Or that it will take a lot longer.

But since plenty of people alive today may live to be 100, it's likely that some of us will be around when Robots become like the smarter ones are in today's science fiction, and we will have them as neighbours and friends and servants. It's a scenario we need to start thinking about.

In novelist Marge Piercy's 1991 book *He, She and It* one of the rules of the society is that no-one is permitted to make a Robot that looks like a human being. You can use all the Artificial Intelligence you want to make machines, but machines with bodies like ours – 'humanoids' – are banned. Of course, in the story, someone breaks the rules, and if you want to know what happens next you will have to read the book.

On the other hand, as the 2013 film *Her* illustrated, perhaps someone could even develop a romantic relationship with a disembodied voice! There are all sorts of clues that make us respond personally to a machine, and physical presence may not be necessary for there to be a lot of confusion. Yet as we know, we can feel close to fictional characters, including those we read about and have never seen on the screen. There's something inside each of us – something that comes with the image of God – that is constantly looking for relationships of one kind or another, and there's no doubt that just as we form them with pets we shall form them with sympathetic machines.

There's a new world awaiting us, not so many years ahead, when Robots will be a major part of our lives. Whether or not they look like us, some of them will be companions, or assistants, or valets, or advisors, with whom we build something not too different from a human relationship.

But if we want a simple way to understand what that will be like, and how we should handle it, we should see these more personal Robots as like pets. Smart pets, pets whom we think of affectionately. But however smart and however affectionate are our pets, we would never suggest that they bear the image of God.

TWO FUTURES?

We face many puzzles as we look into the future.

On the one hand we know that technology developments driven by Moore's Law are racing ahead. People like inventor Ray Kurzweil tend to focus there, and expect the tech progress to shape everything else.

On the other hand, we don't know how people are going to respond – as individuals, and also as governments. People and government can be much influenced by 'opinion leaders' – women and men they respect and who are ahead of the game in their thinking.

It's because people like Bill Gates and Larry Summers have begun to raise concerns about Robots taking jobs that the issue is starting to be taken seriously. It's because Lord Rees and Stephen Hawking and tech luminaries like Elon Musk signed on to the letter about the future of AI that many other people now realise we may one day face big problems.

While technology tends to push events forward, we need to reflect on examples like that of 'genetically-modified' food. Europeans rejected what seemed in the mid-1990s to be obviously the future of agriculture.

The bigger the issue, the bigger we may expect to find the role of people and their opinions.

So perhaps nothing is, in human terms, inevitable! And even if the technology does drive ahead, how humans and their governments respond will shape what happens to us and our families and our neighbours.

Here are two samples of different ways things could go. They assume the same kind of technology developments. The idea is not to look too far – not as far as 2045, whether or not the 'Singularity' awaits us that year as Kurzweil has predicted! But let's go for 2040. Just over two decades from now. A child born in 2017 will just have left university and, perhaps, begun a first job. If you're 21 now, you will be 44. If you're in your early 40s and in the prime of life, you will be about to retire, if people still do then. If you're 65 you'll be 88 – what was once a great age attained by few, though now there are many of us who hope still to be around and leading busy lives!

Let's look ahead to 2040. I have imagined two speeches given by the Prime Minister of the day at party conference.

1. ROBOTS RULE!
THE PRIME MINISTER

The past 20 years have seen amazing developments in technology! And the UK remains one of the top global players.

The economy is booming, the FTSE keeps going up, and the attempts by crazy Luddites to slow everything down have got them nowhere except prison.

Let me summarise the main achievements of this Government so far.

First, the drop-off in jobs continues as Robot systems continue to replace humans in a wide variety of occupations. So the

Chancellor has re-set the model of 'full employment'. This should not be controversial as the previous government lowered it too. They set it at 70 per cent, though they failed to meet their targets year after year. Back in the pre-Robot days of the twentieth century of course it was around 95 per cent of the able-bodied job-seeking population. We have decided to cut it to 50 per cent. And we are glad to announce that we are hitting our target. At least half the people wishing for work now have jobs. I hope we have their votes.

Second, we have moved to lower the state pension age yet again – this time to 50. Efforts to raise it earlier in the century now look very foolish, and show how back then governments were still locked into a pre-Robotic worldview.

Even though the pension payments are lower, this move is proving very popular. The 50 Bonus Plus State Pension is set at 10 per cent higher than unemployment pay, so it's a guaranteed winner. And, as recommended by industry leaders, we have combined it with the reversal of another short-sighted policy from a generation ago. When they hit that age, most people will be required to retire. Plainly some occupations are exempt, and companies need to phase this policy in over time. But there are so many advantages, not least that we can now get more energetic and tech-savvy younger workers to replace their parents. And, of course, they come a lot cheaper!

The phasing out of occupational pension schemes, all of which will conclude by the end of this Parliament, is already saving our vibrant British companies good money they can better deploy to investment in technology and competition against the Asian economies. Of course, the savings to government are enormous as feather-bedded civil service pensions are also ending.

Third, we have decided to continue providing Free Basic Food, one of the best ideas of the previous government. So falling employment is no longer a problem for families, when you take it together with our Free Basic Housing policy. The new buildings going up outside our cities to provide this housing are almost entirely Robot-constructed, which is keeping our costs down, and Robotic surveillance and policing means these new high-rise satellite communities are an economic proposition for us and a safe one for you. Every one of these mega-blocks comes complete with a Free Basic Food store as well as a school, Robot security centre with local prison cells, and of course Free Basic Entertainment is part of the package as well as NHS-Basic.

Meanwhile, government revenues continue to rise, and the greatly reduced costs to businesses now operating with only handfuls of staff mean that corporation taxes more than make up for lost income taxes and National Insurance payments from our diminishing workforce. We're also continuing to cut tax rates for the growing number of very wealthy people whom we wish to encourage to live and invest here rather than in Europe.

This Government remains careful to set our tax policies to encourage Robot innovation at every turn. The efforts by the previous administration to encourage innovation while pushing their foolish 'pro-jobs' agenda are now entirely discredited. You have to make a choice. The British people made theirs, and we have made ours. Just as with the Brexit decision of a quarter-century ago, we see this last election as a decisive turning-point for the British people, and will not permit the Robot issue to be re-opened – whatever we need to do to make sure that happens.

So we have accepted the recommendation of the Royal Commission on the Future, supported by leaders of industry, to restrict voting from the next General Election to people with jobs, so old folk and the unemployed won't get in the way of progress – and it's progress they all need too, so we can support them. We are also assessing the recommendation that we add parliamentary constituencies for companies based on their human-equivalent Robot staffs, which of course has precedent in the old constituencies we used to have for the universities. There will be no return to failed policies and sentimentality.

And I will add a word to the neo-Luddite groups that have now turned to terrorism after their mass demos failed to put the clock back. The assassination of several industry leaders and the awful kidnap and torture of our Minister for Robotisation will get you nowhere. Opinion polls suggesting popular sympathy for your efforts are misleading, and are now happily behind us as the Terror Opinion Poll Order ensures such efforts cannot any more be given the oxygen of publicity.

Looking ahead, we see our policy of encouraging the substitution of capital investment for labour as both ending the opportunity for unions to cripple public services, and funding the tax base we need to enable the Free Basics programmes. Free Basic Food, and Housing, and Entertainment and NHS-Basic.

What the long-term unemployed and those enjoying the blessings of early retirement do with their time is up to them. They are lucky that this nation is sufficiently prosperous to provide for their Basic Needs.

2. PEOPLE POWER!
THE PRIME MINISTER

The grave crises from which we are slowly recovering have given this Government both motive and opportunity to do some very creative thinking about the future.

The 'winter of discontent' of two years ago is now mercifully behind us. The riots across our post-industrial heartland are over for now, and we hope for ever. The kidnapping and YouTube murder of two leading industrialists who had shifted their entire enterprises to Robot production, and a government minister also, have finally begun to recede in our memories, and the leaders of the Ned Ludd Faction are in jail. And we have begun to tackle the spreading poverty and drug abuse that have blighted once prosperous areas as idleness has spread like wildfire across the new Rust Belts of the UK.

As you know, this Government of National Unity was propelled to power by the resignation of the Prime Minister – after her car was overturned and set on fire by rioters in central London. With that violence, we were all fearful of what a General Election might bring. So, instead, we explored avoiding an election with a re-alignment in Parliament. We discovered we could bring together all those willing to fight the disastrous influence on our country of those across all our parties who had sided with the Robot Coalition. Our Technology-for-Humans Movement is seeking to build as broad a consensus as we can achieve.

So while we welcome the advance of technology, we're taken a clear decision that people are always going to come first in this country. The Chancellor's announcement of the Human Preference Tax has made that unambiguous. Every human job converted into a Robot job will now cost the companies concerned more than it will save them – unless they can create the equivalent of 1.5 new jobs elsewhere. And the tax breaks available for companies who do create new human jobs are very generous indeed.

Unlike in many developed nations, our manifesto pledge to be 'Pro-tech, Pro-people' has maintained jobs levels. While they are not the same as the old 'full employment' model of the twentieth century they offer work of some kind to every adult under 60 who wants it. By taxing Robot use in a way that tracks likely company savings, our policy has been designed to give companies a constant choice between machines and humans that is designed both to keep citizens in jobs and encourage creative thinking as well as technology use by managements. And tax incentives are also enabling the creation of large numbers of non-traditional jobs as companies branch into what we used to see as the voluntary sector, and offer the structure of job-type experiences across a wide range of community and cultural projects.

And let me say, we've never ceased to be grateful to Sir Bill Gates, who has graciously accepted an honorary knighthood from the King, and we send him greetings as his 85th birthday comes in a few weeks. While Microsoft, the company he founded, is now forgotten, it once dominated the old-style computing of our parents' generation. But this party has taken very seriously what he said way back in 2015: that governments would need to 'plead' with companies to hire people rather than use Robot machines. Our Robot policy has been aligned precisely with his wisdom, and our tax policies adjusted appropriately. As the Governor of the Bank of England recently said, the board of a public company and its executives need always to have a choice, a free choice: whether to join the Human Company Partnership – with the tax benefits we have provided – or to opt for Robotic alternatives.

Our decision to introduce a Universal Basic Income to replace most social benefits – an idea proposed back in the twentieth century and embraced both by the right and the left – has provided a further level of security to families as the march of technology continues. We are grateful that those companies that have shifted from labour to machines have been so profitable that the exchequer can basically spend the money they are saving by using Robots in place of workers on paying those ex-workers. Whether we call them ex-workers, or the newly early retired, or as John Maynard Keynes, our greatest economist, once famously predicted, the 'new leisured class' – in any case we are paying them. But they are being paid ultimately by the Robots that took their jobs.

Now that everyone has the equivalent of a 'pension', whether they are in work or not, the big question we face is what people are going to do with all that spare time...

And that will be the basis of our manifesto for the next election.

So I am now introducing The Leisure State as our strapline for Britain – and as the agenda of the Royal Commission we are establishing to re-frame the British experience in a world in which retirement is earlier and, of course, people are living longer. As the nation that invented the charity – well, we also invented the joint stock company! – we are perfectly placed to lead the world in our novel approach to the 'work-life balance' for the 2040s.

There's no question that re-distributing the fruits of our industry is central to the concerns of the electorate, and we welcome the move by key business leaders to acknowledge this and embrace it as the partner of the decline of the labour element in the creation of value.

So join me as we look ahead to the future of Britain as The Leisure State!

The UK is determined to find a middle way between the situations we see in the United States, where unemployment and poverty and violence are spreading fast, and the situation in parts of Europe, where innovative technologies have simply been banned. 'Frankenstein technology' is a dangerous label to use. In other European countries, the high social wage has enabled the large numbers now without work to pay their bills but has not given their lives meaning and purpose.

So our Government's policy is to encourage the smartest of our business leaders to welcome technology while also finding inventive ways to employ people.

But at the same time we can see the writing on the wall. Robots will keep getting cheaper and smarter. While we hope to avoid the calamity that has hit parts of the developing world, where the collapse of employment has taken place with little social safety-net and where violence and disorder have been much worse than here, we know that there will be less traditional work to do every year.

So we are launching three new initiatives that we hope will lead to a welcome future for all of our people. What John Maynard Keynes, our greatest economist, called the 'new leisured class'

– ordinary people who find they don't have full-time work – need to be prepared for a life that is very different from that of their grandparents.

The first initiative is to reform our education system so that its priority will be to teach life skills, even before work skills. While there was a big push in recent years to get everyone ready for the technology jobs, we have realised late in the day that these are actually the easiest jobs to turn over to the Robots. People need creative skills, and these skills teach us how to live as well as how to work. So I applaud the Education Secretary's recent speech on The Two Rs: Why Reading and Writing come before Arithmetic.

We have begun to re-vamp our education system so that preparation for 'work' and for leisure now take equal place among its goals. The fresh focus on the 'humanities' – with wide reading as the core of the curriculum, and a shift away from science, technology, engineering and maths for all but the most gifted pupils, the development of the 'creativity central' track for everyone – these moves will better prepare everyone and should prove successful in cultivating the creative genes of those who end up in business and technology.

Second, we are revitalising National Service. All our young people will spend the first three years after graduating working together for us, in all kinds of civil and well as military jobs. And if six months after leaving they still can't find work, they can come back and work for the country.

Third, we are beginning a huge effort to be called the New Festival of Britain – reviving the name of a special event from way back – so as a nation we engage in constant celebration of our people, our human nature, and the wonderful creativity that flows from it. Theatre, the visual arts, literature of many kinds – not only will we be putting on performances all over the country but every citizen will have the right to classes and courses and skilled teaching in everything from writing fiction to making pottery to handicrafts and sewing to cookery. We plan to celebrate our human experience, and while we all love our Robots we want to make sure we love the human factor even more.

What's more, we've decided that having Robot machines that look like people – the 'humanoids' – is confusing, especially to children and older people, whom they often look after. So under the Robot (Suppression of Humanoids) Order 2040, companies have two years to replace all such machines with new models that no longer imitate human looks or human speech. Existing machines will then be destroyed, and there is talk of a national holiday in two years' time when we will destroy some of them together. There will then be a term of five years' imprisonment for any attempt to make another one.

And it's a spiritual issue as much as a human one. I know that most of you long since stopped going to church, and that the establishment of the Church of England may seem out-of-date. But I also know that the Robot crisis has brought us together around the need to rediscover what it means to be human, and that our long religious tradition has been the source of our nation's spiritual vision and relationships and community. We aren't all Christians. I am not. But some of my most impressive friends are, and most of the great leaders of our nation in time past have been. I also spoke last week with the leaders of our Jewish and Islamic and Buddhist and other religious communities, and each of them endorsed enthusiastically what I am going to do next.

I have done something no party leader has ever done before, and I have asked the Archbishop of Canterbury to take over and finish my speech. We aren't all Christians. We certainly aren't all Anglicans. But we are one nation, and for many hundreds of years it has been an Archbishop of Canterbury who has crowned our sovereigns.

So in this time of crisis, as we look to recovery and a human future for this nation, I call on the Archbishop. Your Grace.

THE ARCHBISHOP

Madam Prime Minister, My Lords, Ladies, Gentlemen, and I will add a greeting to Madam Chief Rabbi, and my good friend the Mullah of the Central London Mosque.

When God made us, He gave us three things. First, He made us in His image – so every human being is unique and special. Every choice we make about technology must also be a choice about how special humans are. Second, He commanded us to have dominion over the earth – be fruitful, multiply, fill the earth and subdue it. So we are right to make technologies and the Robot world in which we live is a world we have made under God. Third, He made us accountable. We are stewards. We are not here for our own sake, we are here for his. So how we put these pieces together – how we serve God as image-bearing stewards in a world of Robots – is the key question for us all.

As you may know, we have faced this in the Church. The General Synod of the Church of England decided last year that we would not recognise Robots taking on the roles and duties of pastors and priests. I know that other traditions – Christians especially but also among more liberal Muslims and Reform Jews – now have preachers and counsellors who are not members of our species. We respect those decisions, and the human programmers who have prepared them to do their work.

But after the riots and killings and the desolation of so many of our communities we have decided to back this Government of National Unity and have committed to doing all we can to support the Prime Minister's programme, and her three key initiatives.

There are several ways in which we can look at Christianity. From one point of view it's a religion, like other religions. It is different, in so many ways, but we rejoice that like the other 'Abrahamic' faiths, Judaism and Islam, we celebrate the fact that we humans are made in the image of God. And, with our Hindu and Buddhist friends, we celebrate human life.

From another perspective, we Christians – if I may speak for the Church of England and also the wider Christian community – have been the dominant religion of the West, and therefore of the modern world. While wicked things have been done in the name of the nations that colonised though in some ways civilised the world, and the UK chief was among them, we cherish our commitment to the dignity of every individual, and the uniqueness of those who bear the image of God.

But from another, all we want to know is: What would Jesus do? What would he do if the fishermen from whom he drew most of his disciples had no jobs? What would he do if the woman at the well was a humanoid Robot? What if, what if?

What we do know is this. The Apostle Paul – for those who don't know, the greatest of the early Christian teachers, one of the most exalted rabbis of his day who decided to follow Christ – wrote this:

> 'The Son is the image of the invisible God, the firstborn over all creation. For in him all things were created: things in heaven and on earth, visible and invisible, whether thrones or powers or rulers or authorities; all things have been created through him and for him. He is before all things, and in him all things hold together.'

Our God has not been taken by surprise by these developments. He gave us dominion, and we have exercised it, and it is up to us to forge a human future for real human people in this world where the potential of Robots to help us is as great as their potential to hinder us.

So we throw our weight behind the Prime Minister's initiative. We embrace the Technology-for-Humans Movement. We bear God's image – so we can't deny the technology that comes from our having 'dominion.' Nor the specialness of human nature that most reflects God's image. Nor the stewardship that binds these two together – and shapes our commitment to him and the human future in 2040.

Go back to your families, my friends. Hug your partners and your children, and greet your neighbours. Decommission your humanoids, and re-set your smart voice systems so they sound like the machines they really are.

We're here on Planet Earth – and on Mars and the other celestial bodies where explorers have landed – as humans, mere humans, mortal creatures without many of the smarts of technology, but humans. Humans who we believe bear God's image. Humans who are frail, emotional, often sad, well aware of our powerlessness, but committed to Jesus. Jesus, who created the universe. Jesus, who then became a man, and then 'set his face to go toward Jerusalem'. Jesus, who having made his way there embraced death, even death on a cross. For our sake, and for our salvation.

The Christian vision, my friends, believe it or not, like it or not, is of God himself. He made the Big Bang and all that has followed, and he gave us an awe-inspiring responsibility for this world and all it contains. That means the environment, but it also means science and technology. And at its heart is our need to value people above everything else.

That doesn't dictate any special political programme of the right or the left. The right has generally focused on personal responsibility, and the left on our responsibility for each other, and as Christians we embrace both of these. But at the end of the day people are at the heart of God's purposes, and the more machines we build the more important it is for us keep our eyes on the needs and opportunities of people.

As Robots develop by leaps and bounds, and as it is harder for us to find rewarding work left over for people, we find ourselves on the brink of something very like Hell, or perhaps very like Heaven. Will most people end up poor and unoccupied? Or will they end up reasonably well-off and with useful and interesting things to do with their time?

We don't know if work is really going to disappear, but if it does we need to be ready. So I'm happy to offer my support and the support of Christians and I believe people of other faiths to the human-centred policies of this new, hopeful, united government.

God bless you, Prime Minister.

QUESTIONS TO THINK ABOUT

What's your idea of how things are likely to go over the next 25 years? Do we have much choice – this chapter assumes we do have some – or is the role that technology will play in our futures basically inevitable?

APPENDIX

BOOKS TO READ AND FILMS TO WATCH

If you or your church want to think more about Robots and what they could mean, the simplest thing to do is read some science fiction books and watch some of the films. And then think and talk about what you have been reading and watching. We've discussed several provocative films, including *Bicentennial Man, Be Right Back, Westworld* (the original film and the new TV series), and *AI: Artificial Intelligence*.

We have also mentioned several others – including *The Sixth Day, Her, S1M0NE*, and the *Terminator* series. Other titles that raise some of these questions include *Blade Runner*, an earlier film that is very much on topic, and *I, Robot*.

I have two other books relevant to this discussion. In *How to be a Christian in a Brave New World*, I teamed up with Joni Eareckson Tada to take a look at all kinds of science and technology challenges (Grand Rapids, Mich.: Zondervan, 2006). My new book *Will Robots Take Your Job? A Plea for Consensus* has been published this year (Cambridge: Polity Press, 2017).

For those who want a serious theological read, the best thing out there (there isn't much) is Noreen Herzfeld's *In Our Image. Artificial Intelligence and the Human Spirit* (Minneapolis, Minn.: Fortress Press, 2002). Professor Herzfeld is an expert on Artificial Intelligence who because she is a thoughtful Christian decided to take time off to do a Ph.D. in theology! This short book is the result. It was well worth her effort, and in case you want to go further includes professional bibliographies in both fields.

If you want a sobering look at how Robots could impact jobs, the 2013 report from economists Carl Benedikt Frey and Michael A. Osborne *The Future of Employment: How Susceptible are Jobs to Computerisation?*, Oxford: Oxford Martin School, is the place to start.

There really were 'Luddites,' although it's doubtful there was ever anyone called 'Ned Ludd' in charge. There's a great book that goes into all the details – Malcolm Thomis, *The Luddites. Machine-Breaking in Regency England,* Newton Abbott: David and Charles, 1970.

And here is Keynes' famous essay that peers into the Robotic future: John Maynard Keynes: 'Economic Possibilities for our Grandchildren,' in *Essays in Persuasion,* London: Macmillan, 1931.

ENDNOTES

[1] scientificamerican.com/article/cyborg-10/

[2] businessinsider.com/japan-developing-carebots-for-elderly-care-2015-11

[3] gulf-times.com/story/399901/Japan-works-up-robots-to-assist-care-workers

[4] businessinsider.com/robotic-cat-is-the-perfect-pet-for-seniors-2015-11; parorobots.com/

[5] theguardian.com/society/2014/jul/08/paro-robot-seal-dementia-patients-nhs-japan

[6] economist.com/news/business/21674779-once-regarded-safe-havens-professions-are-now-eye-storm-professor-dr-robot

[7] thememo.com/2016/11/29/ibm-watson-healthcare-app-ibm-watson-hospital-ibm-watson-cancer/

[8] engadget.com/2016/11/30/in-ibms-future-watson-will-be-your-doctor/

[9] techcrunch.com/2016/05/25/affectiva-raises-14-million-to-bring-apps-robots-emotional-intelligence

[10] psychologytoday.com/blog/media-spotlight/201411/the-rise-the-robot-therapist

[11] nytimes.com/2015/09/20/magazine/barbie-wants-to-get-to-know-your-child.html

[12] blogs.scientificamerican.com/guest-blog/why-2016-could-be-a-watershed-year-for-emotional-intelligence-in-machines/

[13] Anthony Painter RSA blog December 16, 2015

[14] scientificamerican.com/article/intelligent-robots-must-uphold-human-rights1/

GLOSSARY

ALGORITHM

A computer programme that enables decisions to be made. Think about how you make a decision, for example whether you will go out tonight; computers are programmed to plug in various factors – am I hungry? can I afford it? would I miss something on TV? and so on – to copy how humans make decisions.

ARTIFICIAL INTELLIGENCE

Computer programmes are getting more and more complicated. We were pretty amazed in the 1970s by pocket calculators and the first 'personal' computers. Now we call banks and credit card companies and chat with machines. Every year things get more sophisticated. There's much disagreement as to whether in the foreseeable future there will be AI that is all-round smart, just like people – what's called General AI. But even if not, much that we do will soon be capable of being done by machines powered by computer chips.

BASIC INCOME

The idea that everyone should be paid by the state, from the age of 18 or so – a kind of state pension. This would cushion people who lose their jobs or only work part-time because Robots take over so many jobs. Interestingly, it's had support from big thinkers on the right as well as the left.

CYBORG

Cyborg is a made-up word for 'cybernetic organism', and cybernetic in this case basically means Robot. So it's a human-Robot mix. Some people say, if you're wearing glasses you're already a cyborg, so what's the problem? Others would argue it's much more serious when our brains start communicating through computers and being made smarter by them (it's already happening).

GPS

Global Positioning System. The satellite-based system, developed by the United States military and then opened up for everyone, that enables us to know where we are by electronic means – so Google maps and every app that notes our location depends on GPS.

IMAGE OF GOD

The core Biblical idea that human beings reflect God. To be human is to be a time-space 'model' of God himself. So he can indeed be our Father, and Jesus our Brother.

INCARNATION

Christians believe that Jesus was God himself born in human form and of a human mother. This idea seems strange to outsiders, and to the greatest Christian thinkers it has proved a mind-bending thought. But it imparts to human nature an extraordinary significance. This isn't just our nature. It was made by God, and then actually taken by him. God became 'one of us'.

INTERNET OF THINGS

This odd term refers to the way in which the internet is now connecting pretty much everything in 'real life'. Lamp-posts, electricity meters, cars, farm machinery – all these things and billions more are getting onto the internet. It is sometimes called the Internet of Everything or the Industrial Internet, and focused in ideas like Smart Cities and the Smart House, it's a fancy way of referring to what we used to call Ubiquitous Computing – computing everywhere.

LUDDITE

Back in the early nineteenth century when machines were being introduced to replace human workers – especially in cotton manufacture – gangs of men went around 'machine-breaking', as it was called, in the north of England. They have ended up being called Luddites, after General Ned Ludd. There probably never was such a person, but his name was quoted at the time and ever since Luddite has meant someone who is anti-technology and wants to wreck it (though not necessarily by terrorist means). The government of the day sent in the army, many people were arrested, and some were hanged. It's become an easy term to use to attack someone who raises serious questions about technology.

MOORE'S LAW

Gordon Moore ran Intel – the 'Intel inside' company that still makes the top-ranked computer chips. Back in the early days he reckoned that every year or two they can cram twice as many circuits on a chip. That's why your mobile phone and laptop keep getting faster – and also cheaper. His 'law' (which has been re-stated once or twice, but not much changed) has worked for more than 30 years.

ROBOTS

I've used the word in this book in a very wide sense, to mean any high-level use of Artificial Intelligence to do something that previously only people could do. Some people use it much more narrowly to mean machines like factory robots, or even only those that look like people

SINGULARITY

This word has various meanings, but it's been given a special one by sci-fi writer Vernor Vinge, and inventor Ray Kurzweil – the point in the future (Kurzweil says it will be in 2045) when machines become smarter than we are, and take over. Others think this will be far, far in the future. Others again, that it will never happen. Elon Musk, Bill Gates, and Stephen Hawking signed a letter sounding a warning that 'rogue AI' could take over.

TECHNOLOGICAL UNEMPLOYMENT

John Maynard Keynes was the most influential economist of the last century. He coined the term 'technological unemployment' for the loss of jobs that machines could cause. He speculated that in the future (he was writing in 1929) the work week might cut by more than half. While this would be wonderful at one level, what would people do with their time?

TURING TEST

Alan Turing was the top computer scientist of the mid-twentieth century, famous for his role at Bletchley Park, where the British Government centre broke German codes in the Second World War. The Turing Test is to tell if you are talking to someone or what we would now call a bot. Obviously, as computers get smarter it gets harder to know.

INDEX

'Artificial intelligence is the future, not only for Russia, but for all humankind. It comes with colossal opportunities, but also threats that are difficult to predict. Whoever becomes the leader in this sphere will become the ruler of the world.' **President Putin**